Everyday Mathematics®

The University of Chicago School Mathematics Project

STUDENT MATH JOURNAL

VOLUME 2

Mc
Graw
Hill
Education

The University of Chicago School Mathematics Project

Max Bell, Director, *Everyday Mathematics* First Edition; James McBride, Director, *Everyday Mathematics* Second Edition; Andy Isaacs, Director, *Everyday Mathematics* Third, CCSS, and Fourth Editions; Amy Dillard, Associate Director, *Everyday Mathematics* Third Edition; Rachel Malpass McCall, Associate Director, *Everyday Mathematics* CCSS and Fourth Editions; Mary Ellen Dairyko, Associate Director, *Everyday Mathematics* Fourth Edition

Authors
Robert Balfanz*, Max Bell, John Bretzlauf, Sarah R. Burns**, William Carroll*, Amy Dillard, Robert Hartfield, Andy Isaacs, James McBride, Kathleen Pitvorec, Denise A. Porter‡, Peter Saecker, Noreen Winningham†

*First Edition only
** Fourth Edition only
†Third Edition only
‡Common Core State Standards Edition only

Fourth Edition Grade 5 Team Leader
Sarah R. Burns

Writers
Melanie S. Arazy, Rosalie A. DeFino, Allison M. Greer, Kathryn M. Rich, Linda M. Sims

Open Response Team
Catherine R. Kelso, Leader; Emily Korzynski

Differentiation Team
Ava Belisle-Chatterjee, Leader; Martin Gartzman, Barbara Molina, Anne Sommers

Digital Development Team
Carla Agard-Strickland, Leader; John Benson, Gregory Berns-Leone, Juan Camilo Acevedo

Virtual Learning Community
Meg Schleppenbach Bates, Cheryl G. Moran, Margaret Sharkey

Technical Art
Diana Barrie, Senior Artist; Cherry Inthalangsy

UCSMP Editorial
Don Reneau, Senior Editor; Rachel Jacobs, Elizabeth Olin, Kristen Pasmore, Loren Santow

Field Test Coordination
Denise A. Porter, Angela Schieffer, Amanda Zimolzak

Field Test Teachers
Diane Bloom, Margaret Condit, Barbara Egofske, Howard Gartzman, Douglas D. Hassett, Aubrey Ignace, Amy Jarrett-Clancy, Heather L. Johnson, Jennifer Kahlenberg, Deborah Laskey, Jennie Magiera, Sara Matson, Stephanie Milzenmacher, Sunmin Park, Justin F. Rees, Toi Smith

Contributors
John Benson, Jeanne Di Domenico, James Flanders, Fran Goldenberg, Lila K. S. Goldstein, Deborah Arron Leslie, Sheila Sconiers, Sandra Vitantonio, Penny Williams

Center for Elementary Mathematics and Science Education Administration
Martin Gartzman, Executive Director; Meri B. Fohran, Jose J. Fragoso, Jr., Regina Littleton, Laurie K. Thrasher

External Reviewers
The *Everyday Mathematics* authors gratefully acknowledge the work of the many scholars and teachers who reviewed plans for this edition. All decisions regarding the content and pedagogy of *Everyday Mathematics* were made by the authors and do not necessarily reflect the views of those listed below.

Elizabeth Babcock, California Academy of Sciences; Arthur J. Baroody, University of Illinois at Urbana-Champaign and University of Denver; Dawn Berk, University of Delaware; Diane J. Briars, Pittsburgh, Pennsylvania; Kathryn B. Chval, University of Missouri–Columbia; Kathleen Cramer, University of Minnesota; Ethan Danahy, Tufts University; Tom de Boor, Grunwald Associates; Louis V. DiBello, University of Illinois at Chicago; Corey Drake, Michigan State University; David Foster, Silicon Valley Mathematics Initiative; Funda Gönülateş, Michigan State University; M. Kathleen Heid, Pennsylvania State University; Natalie Jakucyn, Glenbrook South High School, Glenview, IL; Richard G. Kron, University of Chicago; Richard Lehrer, Vanderbilt University; Susan C. Levine, University of Chicago; Lorraine M. Males, University of Nebraska-Lincoln; Dr. George Mehler, Temple University and Central Bucks School District, Pennsylvania; Kenny Huy Nguyen, North Carolina State University; Mark Oreglia, University of Chicago; Sandra Overcash, Virginia Beach City Public Schools, Virginia; Raedy M. Ping, University of Chicago; Kevin L. Polk, Aveniros LLC; Sarah R. Powell, University of Texas at Austin; Janine T. Remillard, University of Pennsylvania; John P. Smith III, Michigan State University; Mary Kay Stein, University of Pittsburgh; Dale Truding, Arlington Heights District 25, Arlington Heights, Illinois; Judith S. Zawojewski, Illinois Institute of Technology

Note
Many people have contributed to the creation of *Everyday Mathematics*. Visit http://everydaymath.uchicago.edu/authors/ for biographical sketches of *Everyday Mathematics 4* staff and copyright pages from earlier editions.

www.everydaymath.com

Send all inquiries to:
McGraw-Hill Education
8787 Orion Place
Columbus, OH 43240

ISBN: 978-0-02-143100-7
MHID: 0-02-143100-0

Printed in the United States of America.

10 11 12 13 14 QSX 23 22 21 20

Contents

Unit 5

Unit 6

Unit 7

Activity Sheets

Math Boxes

1 Rename each fraction as a whole number or mixed number.

a. $\frac{24}{8} =$ _____   b. $\frac{18}{5} =$ _____

c. $\frac{21}{6} =$ _____   d. $\frac{15}{4} =$ _____

e. $\frac{11}{3} =$ _____

SRB
171

2 Write the following decimals using numerals.

a. three and six hundredths = _____

b. twelve and nine thousandths =

_____

c. seventy and one tenth = _____

SRB
117

3 There are 107 girls at hockey camp. The coach is reserving rinks for games. There can only be 12 girls on each rink. How many rinks should the coach reserve?

_____
(number model)

Solution: _____

What does the remainder represent?

_____

SRB
109,
113–114

4 Carlos rode for 2 hours while training for a bicycle race. In the first hour he rode $15\frac{7}{10}$ miles. In the second hour he rode $14\frac{5}{10}$ miles. Which number model would you use to find the total miles Carlos rode in the 2 hours?

Fill in the circle next to the best answer.

○ **A.** $2 * (15\frac{7}{10} + 14\frac{5}{7}) = m$

○ **B.** $15\frac{7}{10} + 14\frac{5}{10} + 2 = m$

○ **C.** $15\frac{7}{10} + 14\frac{5}{10} = m$

SRB
44

5 Write the ordered pairs for each point on the coordinate grid.

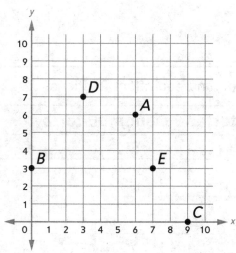

A: (_____ , _____)

B: (_____ , _____)

C: (_____ , _____)

D: (_____ , _____)

E: (_____ , _____)

SRB
275

153

Using Equivalent Fractions to Find Common Denominators

1 Fill in the table by using the information given to help you find equivalent fractions.

SRB
168

Fraction	Multiply Both the Numerator and Denominator by:	Number Sentence	Equivalent Fraction
$\frac{1}{2}$	2	$\frac{(1 * 2)}{(2 * 2)} = \frac{2}{4}$	$\frac{2}{4}$
	3	$\frac{1}{3} \times \frac{2}{3} = \frac{3}{6}$	$\frac{3}{6}$
	5	$\frac{1}{2} \times \frac{5}{5}$	$\frac{5}{10}$
$\frac{2}{5}$	2	$\frac{2}{5} \times \frac{2}{2}$	$\frac{4}{10}$
	3	$\frac{2}{5} \times \frac{3}{3}$	$\frac{6}{15}$
	5	$\frac{2}{5} \times \frac{5}{5}$	$\frac{10}{25}$

What do you notice about the fractions in the gray shaded boxes? _they both_
have tens for denomintor

2 Use the table above to help you solve the following problems.

a. $\frac{1}{2} + \frac{2}{5} =$ ___$\frac{4}{10}$___

b. $\frac{1}{2} - \frac{2}{5} =$ ___$\frac{1}{3}$___

c. Fill in the blank with <, >, or =.

$\frac{1}{2}$ ___$>$___ $\frac{2}{5}$

3 Fill in the table by using the information given to help you find equivalent fractions.

Fraction	Multiply Both the Numerator and Denominator by:				
	2	3	4	5	6
$\frac{2}{3}$	$\frac{4}{6}$	$\frac{6}{9}$	$\frac{8}{12}$	$\frac{10}{15}$	$\frac{12}{30}$
$\frac{1}{6}$	$\frac{2}{12}$		$\frac{4}{24}$		
$\frac{1}{3}$		$\frac{3}{9}$		$\frac{5}{15}$	

Solving Problems Using Common Denominators

SRB
168,
181–182

1 Solve. Use the tables on journal page 154 to find equivalent fractions with a common denominator. Write a number sentence showing the fractions you used.

   a. $\frac{1}{3} - \frac{1}{6} = ?$

   Common denominator: _____

   _____
                             (number sentence)

   b. $\frac{1}{3} + \frac{1}{2} = ?$

   Common denominator: _____

   _____
                             (number sentence)

2 List four equivalent fractions for each fraction given.

   a. $\frac{1}{4} =$ _____, _____, _____, _____   **b.** $\frac{1}{5} =$ _____, _____, _____, _____

3 Make an estimate. Then solve by finding fractions with a common denominator. Use the tables and lists of equivalent fractions above and on journal page 154 to help you. Write a number sentence with a common denominator to summarize each problem.

a. $\frac{1}{2} - \frac{1}{4} = ?$
 	
 Estimate: _____
 	
 Common denominator: _____
 	
 _____
       (number sentence)
 	
 $\frac{1}{2} - \frac{1}{4} =$ _____
	b. $\frac{1}{2} + \frac{1}{6} = ?$
	
	 Estimate: _____
	
	 Common denominator: _____
	
	 _____
	       (number sentence)
	
	 $\frac{1}{2} + \frac{1}{6} =$ _____

c. $\frac{1}{4} + \frac{2}{3} = ?$
 	
 Estimate: _____
 	
 Common denominator: _____
 	
 _____
       (number sentence)
 	
 $\frac{1}{4} + \frac{2}{3} =$ _____
	d. $\frac{1}{4} - \frac{1}{6} = ?$
	
	 Estimate: _____
	
	 Common denominator: _____
	
	 _____
	       (number sentence)
	
	 $\frac{1}{4} - \frac{1}{6} =$ _____

4 Rewrite the fractions as equivalent fractions with a common denominator. Fill in the blank with >, <, or = to make a true number sentence.

   a. $\frac{1}{2}$ _____ $\frac{2}{5}$ Fractions with a common denominator: _____, _____

   b. $\frac{1}{3}$ _____ $\frac{2}{6}$ Fractions with a common denominator: _____, _____

Practicing Common Denominator Strategies

SRB
177, 190

Summarize the strategies discussed in class for finding a common denominator.
Circle the ones that always work.
Put a star next to the strategy that you prefer.

Strategy 1 _____

_____

Strategy 2 _____

_____

Strategy 3 _____

_____

Explain why you prefer the strategy you starred. _____

_____

_____

Use any of these strategies to find common denominators.
Rewrite each pair of fractions using a common denominator. Then solve.

1 a. $\frac{2}{9}$ and $\frac{5}{6}$. Common denominator: ___18___ $\frac{2}{9} = \frac{4}{18}$ $\frac{5}{6} = \frac{15}{18}$

b. $\frac{2}{9} + \frac{5}{6} = \frac{19}{18}$ **c.** $\frac{5}{6} - \frac{2}{9} = \frac{11}{18}$

d. Fill in the blank with >, <, or = : $\frac{2}{9}$ ___<___ $\frac{5}{6}$

2 a. $\frac{3}{4}$ and $\frac{7}{12}$. Common denominator: ___12___ $\frac{3 \times 3}{4 \times 3} = \frac{9}{12}$ $\frac{7}{12} = \frac{7}{12}$

b. $\frac{3}{4} + \frac{7}{12} = \frac{16}{12}$ or $1\frac{4}{12}$ **c.** $\frac{3}{4} - \frac{7}{12} = \frac{2}{12}$

d. Fill in the blank with >, <, or = : $\frac{3}{4}$ ___>___ $\frac{7}{12}$

3 a. $\frac{4}{7}$ and $\frac{1}{2}$. Common denominator: ___3___ $\frac{4}{7} = $ _____ $\frac{1}{2} = $ _____

b. $\frac{4}{7} + \frac{1}{2} = $ _____   **c.** $\frac{4}{7} - \frac{1}{2} = $ _____

d. Fill in the blank with >, <, or = : $\frac{4}{7}$ _____ $\frac{1}{2}$

156

Practicing Common Denominator Strategies (continued)

For each problem, find a common denominator and write which strategy you used. Then write the fractions using a common denominator and solve.

SRB
177, 190

Strategies:

- I listed equivalent fractions.

- I noticed one denominator was a multiple of the other denominator.

- I found a quick common denominator.

4 $\frac{2}{3}$ and $\frac{10}{15}$

    a. Strategy: _____

    b. Fractions with a common denominator: _____ and _____

    c. $\frac{2}{3} + \frac{10}{15} =$ _____
                 d. Fill in the blank with <, >, or =: $\frac{2}{3}$ _____ $\frac{10}{15}$

5 $\frac{1}{4}$ and $\frac{2}{9}$

    a. Strategy: _____

    b. Fractions with a common denominator: _____ and _____

    c. $\frac{1}{4} - \frac{2}{9} =$ _____
                 d. Fill in the blank with <, >, or =: $\frac{1}{4}$ _____ $\frac{2}{9}$

6 $\frac{5}{6}$ and $\frac{3}{4}$

    a. Strategy: _____

    b. Fractions with a common denominator: _____ and _____

    c. $\frac{5}{6} + \frac{3}{4} =$ _____
                 d. $\frac{5}{6} - \frac{3}{4} =$ _____

7 **a.** What other strategy could you have used to find a common denominator in Problem 6?

           _____

           _____

    b. Which strategy do you think is best for this problem? Why?

           _____

           _____

Math Boxes

1 Shade the first grid to represent seven tenths.
Shade the second grid to represent sixty-seven hundredths.

Write >, <, or = to make a true number sentence.

0.7 _____ 0.67

SRB
120–121

2 A marathon is 26 miles, 1,056 feet.
A mile is 5,280 feet. Write a number
model to show the number of feet in a
marathon. Then solve.

_____
(number model)

Answer: _____ feet

SRB
44, 328

3 Michalene is buying sports equipment for
a picnic. She buys a badminton set for
$49.99 and a beanbag-toss set for
$129.99. How much money will Michalene
spend? Make an estimate and solve.

_____
(estimate)

Answer: _____

SRB
128, 130

4 **Writing/Reasoning** How did the grids help you compare the decimals in Problem 1?

_____

_____

_____

_____

SRB
121

Math Boxes

Math Boxes

---

1 Rename each mixed number as an equivalent mixed number with the same denominator.

a. $7\frac{5}{3}$ = _____

b. $4\frac{6}{9}$ = _____

c. $3\frac{7}{8}$ = _____

d. $5\frac{19}{14}$ = _____

SRB
173

---

2 Write each of the following decimals in words:

a. 18.04 = _____

_____

b. 814.017 = _____

_____

SRB
117

---

3 Hannah and her three brothers equally split the cost of a $379 gift for their parents. How much did each sibling pay?

a. Each sibling paid _____ dollars.

b. What did you do with the remainder?
   Rounded the quotient up

   Reported it as a fraction

   Ignored it

SRB
109,
113–114

---

4 Mindy needs $3\frac{1}{3}$ cups of raisins for the trail mix she's making. She only has $1\frac{2}{3}$ cups of raisins in her pantry. How many more cups of raisins does she need? Show your work.

_____
(number model)

Mindy needs an additional _____ cups of raisins.

SRB
178–180,
188

---

5 Plot the following points on the grid. Then connect them in order.

(0, 1) (1, 3) (4, 3) (5, 1) (0, 1)

What shape have you drawn?

_____

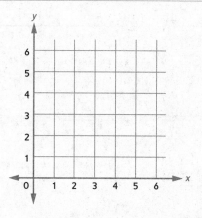

SRB
268, 275

159

Adding Fractions and Mixed Numbers

Math Message Choose the best estimate for each problem. Then write the answer for any problem that you can solve mentally.

Estimate:

1. $\frac{2}{5} + \frac{1}{4} = $ _____

☐ about $\frac{1}{2}$

☐ about 1

☐ about $1\frac{1}{2}$

2. $\frac{1}{9} + \frac{4}{9} + \frac{3}{9} = $ _____

☐ about $\frac{1}{2}$

☐ about 1

☐ about $1\frac{1}{2}$

3. $\frac{3}{4} + \frac{5}{8} = $ _____

☐ a little less than 1

☐ a little more than 1

☐ a little more than 2

4. $2\frac{4}{5}$
   $+ 1\frac{2}{3}$
   _____

☐ about 3

☐ a little less than 4

☐ between 4 and 5

5. $2\frac{1}{16}$
   $+ 5\frac{1}{16}$
   _____

☐ a little less than 7

☐ a little more than 7

☐ about 8

6. $2\frac{7}{8}$
   $+ 1\frac{3}{8}$
   _____

☐ a little more than 3

☐ a little less than 4

☐ a little more than 4

Solving Addition Problems with Fractions and Mixed Numbers

Estimate each sum and then solve. Show your work.

SRB
181, 187,
189–191

1   Estimate: _____

$3\frac{4}{7}$
$+ 4\frac{4}{7}$

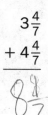

2   Estimate: _____

$6\frac{3}{4}$
$+ \frac{1}{6}$

3   Estimate: _____

$\frac{7}{8}$
$+ \frac{1}{6}$

4   Estimate: _____

$7\frac{2}{3}$
$+ 2\frac{3}{5}$

For each number story, write a number model with an unknown and make an estimate. Then solve the story. Show your work. Record your answer and a summary number model. Use your estimate to check whether your answer makes sense.

5   Mr. Kumar's class ate $6\frac{3}{4}$ pizzas, and Ms. Rinehart's class ate $4\frac{2}{4}$ pizzas. How many pizzas did the two classes eat?

Number model: _____

Estimate: _____

Answer: _____ pizzas   Summary number model: _____

161

⑥ Melanie's superhero costume for the school play requires $1\frac{5}{6}$ yards of green fabric and $\frac{1}{3}$ yard of yellow fabric. How many total yards of fabric are needed for the costume?

SRB
181, 187,
189–191

Number model: _____

Estimate: _____

Answer: _____ yards   Summary number model: _____

⑦ Charlotte ran $5\frac{2}{3}$ miles on Monday and $1\frac{5}{8}$ miles on Tuesday. How many miles did she run in all?

Number model: _____

Estimate: _____

Answer: _____ miles   Summary number model: _____

Subtracting Fractions and Mixed Numbers

Math Message

SRB
173,
181–193

① Estimate: _____

$$4\tfrac{4}{5}$$
$$-1\tfrac{1}{5}$$

$3\tfrac{3}{5}$

② Estimate: _____

$$3\tfrac{1}{3}$$
$$-1\tfrac{2}{3}$$

③ Estimate: _____

$$\tfrac{9}{10} - \tfrac{8}{10} = \tfrac{1}{10}$$

④ Estimate: _____

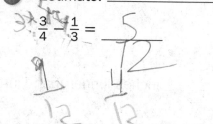

$$\tfrac{3}{4} - \tfrac{1}{3} = \tfrac{5}{12}$$

$\tfrac{1}{12}$ $\tfrac{4}{12}$

⑤ Estimate: _____

$$3\tfrac{1}{3}$$
$$-1\tfrac{1}{2}$$

⑥ Estimate: _____

$$10\tfrac{1}{5}$$
$$-4\tfrac{2}{3}$$

163

Subtraction with Fractions and Mixed Numbers

SRB
173,
181–193

1 Fill in the missing numbers.

a. $5\frac{1}{4} = 4\frac{\square}{4}$

b. $8\frac{7}{9} = \underline{\quad}\frac{16}{9}$

c. $\underline{\quad}\frac{3}{6} = 3\frac{9}{6}$

2 Estimate: _____

$\frac{7}{12} - \frac{3}{8} = \underline{\quad}$

3 Jake has two pet guinea pigs named Fluffy and Scruffy. Fluffy is $8\frac{1}{8}$ inches long. Scruffy is $10\frac{1}{4}$ inches long. How much longer is Scruffy than Fluffy?

Number model: _____

Estimate: _____

Scruffy is _____ inches longer than Fluffy.

4 Rachel is traveling by plane from Chicago to San Diego. The flight will take $4\frac{1}{4}$ hours. The plane took off $1\frac{2}{3}$ hours ago. How much longer will Rachel be on the plane?

Number model: _____

Estimate: _____

Rachel will be on the plane _____ more hours.

5 Estimate: _____

$\begin{array}{r} 4\frac{1}{6} \\ -\ 3\frac{2}{6} \\ \hline \end{array}$

Try This

6 Estimate: _____

$\begin{array}{r} 9 \\ -\ 4\frac{7}{8} \\ \hline \end{array}$

Reviewing Decimals

1. Use the clues to write the mystery number: ____. ____ ____ ____

   SRB
   115–127

   - I have a 6 in the hundredths place.
   - I have a 2 in the tenths place.
   - I have a 9 in the ones place.
   - I have a 1 in the thousandths place.

2. Complete the table below.

Numeral	Words	Value of the Digit 4
3.409	three and four hundred nine thousandths	4 tenths
2.541		
	eighty-four thousandths	
4.06		
	four hundred ten thousandths	

3. Look at the table. How does the place-value position of the 4 affect the value of the 4?

   _____

   _____

4. Write each number in expanded form.

   a. 4.573 _____

   b. 32.081 _____

5. Complete each number sentence with <, >, or =.

   a. 0.816 _____ 0.82   b. 0.8 _____ 0.82

   c. 0.9 _____ 0.095   d. 0.300 _____ 0.30

   e. 0.076 _____ 0.067   f. 0.254 _____ 0.257

6. Put the following decimals in order from least to greatest: 0.82, 0.816, 0.095, 0.9.

   _____, _____, _____, _____

7. Round to complete the table.

Start Number	Rounded to the Nearest Hundredth	Rounded to the Nearest Tenth	Rounded to the Nearest Whole Number
3.409			
4.573			
0.816			

Math Boxes

1 Shade the first grid to represent one tenth.
Shade the second grid to represent ninety-nine thousandths.

Write >, <, or = to make a true number sentence.

0.1 _____ 0.099

SRB
120–121

2 Shawn is 38 years, 24 days old.
How many days has he been alive?
Remember: There are 365 days in a year.
(You don't need to count leap years.)

_____
(estimate)

Answer: _____ days

SRB
76–77,
328

3 Brooke's total at the grocery store is
$73.26. She is paying with four $20 bills.
What will her change be?

_____
(estimate)

Answer: _____

SRB
128,
131–132

4 **Writing/Reasoning** Explain how you could use place value to compare the decimals in Problem 1.

_____

_____

_____

SRB
122–123

166

Solving Fraction-Of Problems

Math Message

1 Use the rule to complete the table.

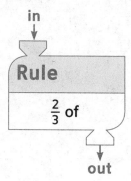

in	out
10	
25	
15	
30	

2 Look at this function machine and table. Compare it to the function machine and table in Problem 1. Talk to a partner about what you notice.

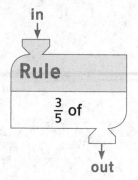

in	out
10	6
25	15
15	9
30	18

---

3 Fill in the missing *out* numbers.

in	out
9	6
24	16
12	
6	

4 Fill in the missing *out* numbers.

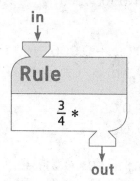

in	out
8	
20	
16	
28	

5 Stella had a carton of 12 eggs. She used $\frac{4}{6}$ of the eggs to bake loaves of bread for a bake sale.

   a. How many eggs are in $\frac{1}{6}$ of the carton? _____ eggs

   b. How many eggs did Stella use for the bread? _____ eggs

   c. Write a multiplication number model for the problem. _____

6 A shoe store had 30 pairs of tennis shoes in stock. One week they had a sale and sold $\frac{8}{10}$ of the shoes. How many pairs of tennis shoes did they sell that week?

Answer: _____ pairs of tennis shoes

_____
(multiplication number model)

167

Math Boxes

---

1 Find a common denominator for $\frac{1}{2}$ and $\frac{1}{3}$. Then solve the problems.

Common denominator: _____

$\frac{1}{2} + \frac{1}{3} =$ _____

$\frac{1}{2} - \frac{1}{3} =$ _____

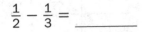
SRB
177,
189–190

---

2 On average, hair grows about 1.25 cm per month. Fingernails grow about 0.3 cm per month. About how much more does hair grow in a month than fingernails?

_____
(estimate)

Hair grows about _____ cm more per month.

SRB
128,
131–132

---

3 Are the expressions below equal to 57.026? Fill in Yes or No for each expression.

A. 50 + 7 + 0 + 0.02 + 0.006
◯ Yes   ◯ No

B. 50 + 7 + 0.2 + 0.06
◯ Yes   ◯ No

C. (5 * 10) + (7 * 1) + (2 * 0.01) + (6 * 0.001)
◯ Yes   ◯ No

SRB
118

---

4 Round each decimal to the nearest tenth.

a. 0.79 _____

b. 3.12 _____

c. 813.47 _____

d. 6.05 _____

SRB
124–127

---

5 Elias wants to finish the last 89 pages of his book in 3 hours. If he reads at the same speed the whole time, how many pages must he read each hour? Report your answer as a mixed number.

_____
(number model)

Elias must read _____ pages each hour.

SRB
44, 108,
113–114

---

6 Estimate and solve.

912 × 87 = ?

_____
(estimate)

912 × 87 = _____

SRB
83, 100,
103–104

Math Boxes

More Function Machines

Math Message

SRB
53–54
195–196

In Problems 1 and 2, use the rules to complete the tables. Then look carefully at your answers and talk with a partner about what you notice.

1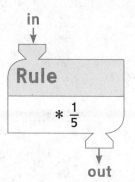

in	out
15	
20	
35	
10	

2

in	out
15	
20	
35	
10	

---

3 Use the rule to fill in the missing *out* numbers.

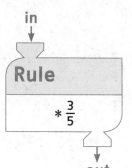

in	out
15	
20	
35	
10	

4 Use the rule to fill in the missing *out* numbers.

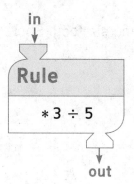

in	out
15	
20	
35	
10	

5 In your own words, describe two strategies you know for multiplying fractions by whole numbers.

a. Strategy 1: Think of a fraction-of problem.

_____

_____

_____

b. Strategy 2: Interpret the fraction as division.

_____

_____

_____

169

Multiplying Whole Numbers by Fractions

1 $13 * \frac{2}{3} = ?$

a. Solve the problem by thinking about $\frac{2}{3}$ of 13. Show your work.

b. Solve the problem by thinking about $\frac{2}{3}$ as $2 \div 3$. Show your work.

$13 * \frac{2}{3} =$ _____

$13 * \frac{2}{3} =$ _____

Solve Problems 2–5 using any strategy. Show your work.

2 $16 * \frac{3}{4} = ?$

3 $20 * \frac{5}{8} = ?$

$16 * \frac{3}{4} =$ _____

$20 * \frac{5}{8} =$ _____

4 Lydia had 10 feet of ribbon. She used $\frac{4}{5}$ of it to tie a bow on a large gift box. How much ribbon did she use?

Number model: _____

5 In a classroom, there are 14 cubbies along a wall. Each cubby is $\frac{5}{6}$ foot wide. How long is the line of cubbies?

Number model: _____

Answer: _____ feet

Answer: _____ feet

Using a Graph to Answer Questions

Oliver is using a garden hose to fill his pool. The table below shows the depth of the water in inches after different numbers of hours.

SRB
55–56,
275

Time in Hours (x)	Water Depth in Inches (y)
0	0
1	6
2	12
3	18
5	30

① Write the data as ordered pairs.

(_____ , _____)

(_____ , _____)

(_____ , _____)

(_____ , _____)

(_____ , _____)

② Graph the points on the grid. Draw a line to connect the points.

Use the graph to help you solve Problems 3–5.

③ How many inches deep was the water . . .

a. after 4 hours? _____ inches

b. after 6 hours? _____ inches

c. after $2\frac{1}{2}$ hours? _____ inches

④ After how many hours was the water . . .

a. 42 inches deep? _____ hours

b. 27 inches deep? _____ hours

c. 3 inches deep? _____ hour

⑤ a. Oliver wants the water to be $3\frac{1}{2}$ feet deep. After how many hours should he turn off the water? _____ hours

b. Explain how you solved Problem 5a.

_____

_____

_____

_____

_____

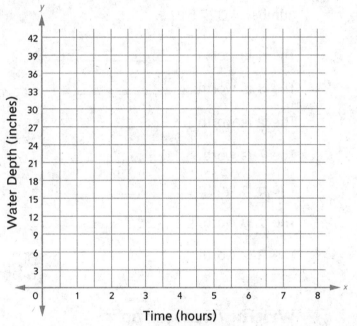

171

Math Boxes

---

1 $4,072 \div 39 = ?$

_____
(estimate)

$4,072 \div 39 \rightarrow$ _____

SRB
84,
109–110

---

2 Round each amount to the nearest dollar.

a. $4.99 $_____

b. $27.25 $_____

c. $119.53 $_____

SRB
124–127

---

3 Write the value of each digit in the number 4,327.519.

The 4 is worth _____.

The 3 is worth _____.

The 2 is worth _____.

The 7 is worth _____.

The 5 is worth _____.

The 1 is worth _____.

The 9 is worth _____.

SRB
118–120

---

4 Destiny is making two casseroles. One recipe uses $\frac{3}{4}$ lb mushrooms. The other uses $\frac{1}{10}$ lb mushrooms. How many pounds of mushrooms does she need?

_____
(number model)

Destiny needs _____ of mushrooms.

SRB
178–180,
189–190

---

5 **Writing/Reasoning**
Illustrate your solution to Problem 1 with an area model.

Area (Dividend): _____

Length (Divisor): _____

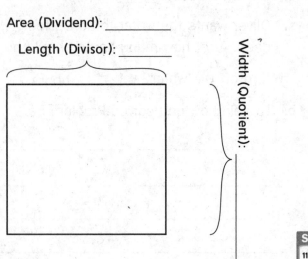

Width (Quotient):

SRB
111–112

Paper-Folding Problems

1 Think of the rectangles below as sheets of paper. Draw fold lines and shade to show how you found $\frac{1}{3}$ of $\frac{1}{2}$ and $\frac{2}{3}$ of $\frac{1}{2}$ while solving Ava's pizza problem.

a. $\frac{1}{3}$ of $\frac{1}{2}$ is _____.

b. $\frac{2}{3}$ of $\frac{1}{2}$ is _____.

2 Fold paper to help you solve this number story:

Carolyn had $\frac{1}{3}$ liter of water. She drank $\frac{3}{4}$ of the water. What part of a liter did she drink?

Think: *What is $\frac{3}{4}$ of $\frac{1}{3}$?*

a. Fold and shade paper to show $\frac{1}{3}$. Then fold and double-shade it to show $\frac{1}{4}$ of $\frac{1}{3}$. Record your work below.

b. Add double-shading to your paper to show $\frac{3}{4}$ of $\frac{1}{3}$. Record your work below.

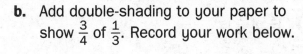

$\frac{1}{4}$ of $\frac{1}{3}$ is _____.

$\frac{3}{4}$ of $\frac{1}{3}$ is _____.

c. What part of a liter of water did Carolyn drink? _____ liter

Solve Problems 3–6 by folding paper. Then use the rectangles to show what you did.

SRB
201

3 $\frac{1}{2}$ of $\frac{1}{4}$ is _____.

4 $\frac{1}{3}$ of $\frac{2}{3}$ is _____.

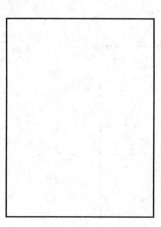

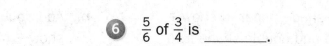

5 $\frac{2}{3}$ of $\frac{1}{4}$ is _____.

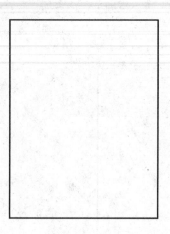

6 $\frac{5}{6}$ of $\frac{3}{4}$ is _____.

1   Find a common denominator for $\frac{2}{3}$ and $\frac{2}{5}$. Then solve the problems.

Common denominator: _____

$\frac{2}{3} + \frac{2}{5} =$ _____

$\frac{2}{3} - \frac{2}{5} =$ _____

SRB
177,
189–190

2   A carpenter is making one shelf that is 37.5 cm long and another that is 54.7 cm long. How much wood does she need?

_____
(estimate)

She needs _____ cm of wood.

SRB
128, 130

3   Which of the following show 318.999 in expanded form?

Fill in the circle next to all that apply.

(A) $(3 * 100) + (1 * 10) + (8 * 1) +$ $(9 * \frac{1}{10}) + (9 * \frac{1}{100}) + (9 * \frac{1}{1,000})$

(B) $(3 * 100) + (1 * 10) + (8 * 1) +$ $(9 * 0.1) + (9 * 0.01) + (9 * 0.001)$

(C) $300 + 10 + 8 + 9 + 0.9 + 0.09$

(D) $300 + 10 + 8 + 0.9 + 0.09$ $+ 0.009$

SRB
118

4   Round each decimal to the nearest hundredth.

a. 0.793 _____

b. 3.125 _____

c. 813.473 _____

d. 6.097 _____

SRB
124–127

5   Cho and three friends made $217 mowing lawns. How much will each friend get when they split the money evenly? Write your answer as a mixed number.

_____
(number model)

Each friend will get _____ dollars.

SRB
44, 108,
113–114

6 $302 * 57 = ?$

_____
(estimate)

$302 * 57 =$ _____

SRB
83, 100,
103–104

Area Models for Fraction Multiplication

1 This diagram shows one way to represent $\frac{2}{3} \times \frac{2}{3}$.

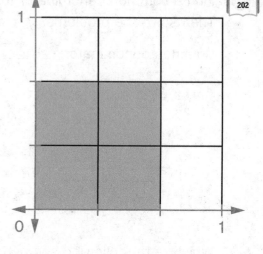

   a. What are the dimensions of the large square?

   _____ unit by _____ unit

   b. What is the area of the large square?

   _____ square unit

   c. What are the dimensions of the shaded rectangle?

   _____ unit by _____ unit

   d. What is the area of the shaded rectangle?

   _____ square unit

   e. Write a multiplication number sentence for the area of the shaded rectangle.

   _____

2 a. Label the blank tick marks on the number lines.

   b. What are the dimensions of the shaded

   rectangle? _____ unit by _____ unit

   c. What is the area of the shaded rectangle?

   Hint: The area of the big square is 1 square unit.

   _____ square unit

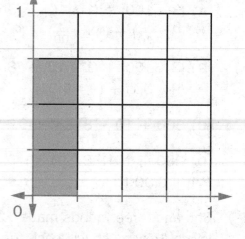

   d. Write a multiplication number sentence for the

   area of the shaded rectangle. _____

3 a. Label the blank tick marks on the number lines.

   b. What are the dimensions of the shaded rectangle?

   _____ unit by _____ unit

   c. What is the area of the shaded rectangle?

   Hint: The area of the big square is 1 square unit.

   _____ square unit

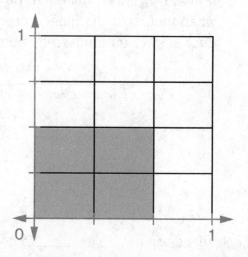

   d. Write a multiplication number sentence for the

   area of the shaded rectangle. _____

Using Area Models to Multiply Fractions

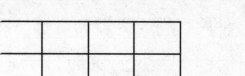

SRB
202

1 Follow these steps using the diagram at the right to help you multiply $\frac{3}{4}$ by $\frac{2}{6}$.

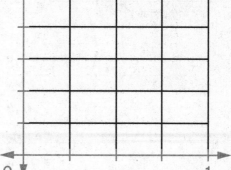

a. Label the blank tick marks on the horizontal number line.

b. Label the blank tick marks on the vertical number line.

c. Shade a rectangle that is _____

unit long and _____ unit wide.

d. Find the area of the shaded rectangle.

_____ square unit

e. Fill in the blank to make a true number sentence: $\frac{3}{4} \times \frac{2}{6} =$ _____

In Problems 2 and 3:

• Label the blank tick marks.

• Shade a rectangle with the given dimensions.

• Find the area of your rectangle.

• Fill in the blank to complete the fraction multiplication number sentence.

2 Dimensions: $\frac{3}{5}$ by $\frac{2}{3}$

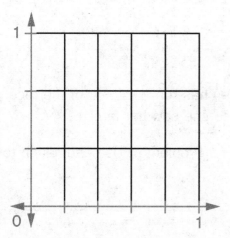

Area: _____ square unit

$\frac{3}{5} \times \frac{2}{3} =$ _____

3 Dimensions: $\frac{5}{6}$ by $\frac{1}{4}$

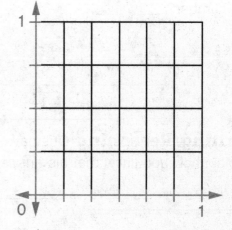

Area: _____ square unit

$\frac{5}{6} \times \frac{1}{4} =$ _____

177

Math Boxes

Math Boxes

1   9,852 ÷ 93 = ?

_____
(estimate)

9,852 ÷ 93 → _____

SRB
84,
109–110

2   Circle the number that shows each decimal rounded to the nearest hundredth.

a. 7.585   7.58 or 7.59

b. 5.004   5.00 or 5.01

c. 23.072   23.07 or 23.08

SRB
124–127

3 **a.** Using only the digits 6, 9, and 1, what is the largest decimal less than 1 that you can write? _____

b. Using only the digits 6, 9, and 1, what is the smallest decimal you can write? _____

c. What is the value of the 6 in each number? _____

SRB
118–120

4   Gwen is $\frac{7}{8}$ of the way through a race. She saw her family cheering when she was $\frac{2}{3}$ of the way done. How much of the race has Gwen run since she saw her family?

_____
(number model)

Gwen has run _____ of the race since she saw her family.

SRB
178–180,
189–190

5 **Writing/Reasoning** Otis gave an answer of $\frac{5}{5}$ for Problem 4. Use estimation to explain how you know that his answer is incorrect.

_____

_____

_____

SRB
181–182

178

Multiplying Fractions

1 In your own words, describe the method for multiplying fractions discovered in class.

SRB
197–198,
202–203

_____

_____

_____

Use the fraction multiplication algorithm described above to solve Problems 2–7.

2 $\frac{1}{2} * \frac{3}{6} =$ _3_
_3_

3 $\frac{2}{3} * \frac{1}{4} =$ _1_

4 $\frac{3}{5} * \frac{1}{6} =$ _____

5 $\frac{3}{4} * \frac{3}{8} =$ _1_
_2_

6 $\frac{2}{5} * \frac{4}{10} =$ _2_
_2_

7 $\frac{7}{9} * \frac{2}{12} =$ _____

8 Choose one of the problems above. Draw an area model for the problem. Explain how it shows that your answer is correct.

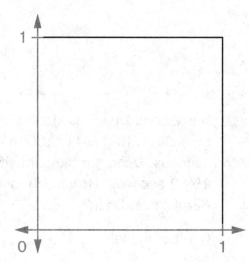

_____

_____

_____

_____

_____

For Problems 9 and 10, write a number model. Then solve.

9 Sheila had $\frac{3}{4}$ pound of blueberries. She used $\frac{1}{3}$ of them in a fruit salad. How many pounds of blueberries did she use?

Number model: _____

Answer: _____ pound

10 The mirror in a dollhouse is $\frac{2}{4}$-inch wide and $\frac{3}{4}$-inch tall. What is the area of the mirror in square inches?

Number model: _____

Answer: _____ square inch

11 Ben tried to solve Problem 9 and got the answer $\frac{4}{7}$. He said, "That can't be right because $\frac{1}{3}$ is less than $\frac{4}{7}$." Do you agree with Ben? Explain.

_____

_____

Solving Decimal Number Stories

For each story, write a number model with a letter for the unknown. Then estimate and solve. **SRB** 44, 128, 130–132

1 Janelle bought milk and bread. Including tax, the milk cost $2.68 and the bread cost $3.42. How much did Janelle spend?

Number model: _____

Estimate: _____

Janelle spent _____.

2 Gene bought a book for $8.79. He paid with a $10 bill. How much change did Gene get?

Number model: _____

Estimate: _____

Gene received _____ in change.

3 Harper and Dean raced each other in gym class. Harper ran 100 m in 14.56 seconds. Dean ran the same distance in 15.12 seconds. How much faster was Harper than Dean?

Number model: _____

Estimate: _____

Harper beat Dean by _____ second(s).

4 Diana is using the Internet for research. A website showed that it took 0.51 second to do her first search, 0.26 second for her second, and 0.28 second for her third. How long did all three searches take?

Number model: _____

Estimate: _____

It took the search engine _____ second(s) to complete all three searches.

5 Explain how you solved Problem 3.

_____

_____

_____

Math Boxes

1 Solve. Show your work.

$$8\frac{5}{6} + 1\frac{1}{7} = \underline{\hspace{1.5cm}}$$

SRB
177, 191

2 The normal temperature of the human body is 98.6°F. Marcus has a fever of 101.8°F. How much higher is his temperature than normal body temperature?

_____
(estimate)

Marcus's temperature is _____ degrees higher than normal.

SRB
128,
131–132

3 Irma is making two types of bows to sell at the craft fair. She wants to make a sample of each type to show her teacher. She needs $\frac{3}{4}$ yard of ribbon for one type and $\frac{3}{8}$ yard of ribbon for the other. How much ribbon does she need in all?

Irma needs _____ yards of ribbon.

SRB
178–180,
189–190

4 Write a number model to represent the story. Then solve.

Alex earns $8 per hour when he babysits. How much will he earn in $\frac{3}{4}$ hour?

_____
(number model)

Alex will earn $_____.

SRB
44, 196,
199–200

5 **Writing/Reasoning** Explain the strategy you used to subtract in Problem 2.

_____

_____

_____

SRB
131–132

Representing Fraction Multiplication

① Draw a picture or fold a piece of paper to help you find $\frac{1}{3}$ of $\frac{2}{5}$. _____

② Explain how your picture or paper folding represents the problem.

■ Math Boxes

1 Write each number using exponential notation.

   a. 10,000,000 = _____

   b. 100,000 = _____

SRB
68

2 Solve.

   8 * 700 = _____

   36,000 = _____ * 40

   320,000 = 800 * _____

   2,000 * _____ = 24,000

   5,000 * 4,000 = _____

SRB
97–98

3 Find the volume of the figure below.

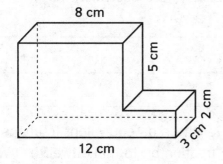

Volume = _____ cm³

SRB
233–234

4 Fill in the blank with <, >, or =.

   $30 * \frac{3}{10}$ _____ 30

   30 _____ 30 * 0.3

   $30 * \frac{3}{10}$ _____ 30 * 0.3

SRB
197–198

5 Place the fractions and mixed numbers on the number line.

$$\frac{7}{3} \quad \frac{1}{4} \quad \frac{3}{2} \quad 2\frac{1}{4}$$

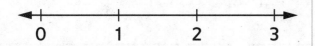

SRB
158–160

6 Look at the number sentences below.

   $6 \times 10^2 = 600$

   $1 \times 10^7 = 10,000,000$

   $9 \times 10^5 = 900,000$

   Explain the pattern in the number of zeros in each product.

   _____

   _____

   _____

SRB
95–96

Explaining the Equivalent Fractions Rule

Math Message

SRB
152, 168,
197–198

1 Fill in the blanks with >, <, or =.

a. $5 * \frac{1}{2}$ _____ 5   **b.** $5 * 2\frac{1}{3}$ _____ 5   **c.** $5 * \frac{5}{4}$ _____ 5

d. $5 * \frac{86}{87}$ _____ 5   **e.** $5 * 1$ _____ 5

---

2 **a.** Write three fractions that are equivalent to 1. _____, _____, _____

b. Use the fractions you wrote in Part a to write three fractions equivalent to $\frac{4}{5}$.

$\frac{4}{5} * \boxed{\frac{3}{3}} = \frac{12}{15}$ _____, _____, _____

c. Explain how you know that the fractions you found in Part b are equivalent to $\frac{4}{5}$.

_____

_____

_____

3 Look at the fractions in the table below. They can all be rewritten as equivalent fractions with a denominator of 12. What would you multiply each fraction by to make twelfths? Write the fraction name for 1 that you would use in each case. Then find the equivalent fraction with a denominator of 12. One row has been completed for you.

Original Fraction	Fraction Name for 1	Equivalent Fraction
$\frac{2}{3}$	$\frac{4}{4}$	$\frac{8}{12}$
$\frac{3}{4}$		
$\frac{5}{6}$		
$\frac{1}{2}$		

4 Pick one fraction name for 1 that you used in Problem 3. How do you know the fraction is equivalent to 1?

_____

_____

Math Boxes

Math Boxes

---

① Fill in the blank with <, >, or =.

a. $\frac{2}{3} \times 17$ _____ 17

b. $\frac{19}{20} \times 13$ _____ 13

c. $\frac{11}{12} \times 52$ _____ $52 \times \frac{11}{12}$

SRB 197–198

---

② Solve. Show your work.

$$4\frac{3}{5}$$
$$-\ \ 2\frac{1}{2}$$

SRB 177, 192–193

---

③ Shade the rectangle to illustrate how you would fold paper to find $\frac{1}{4} * \frac{2}{3}$.

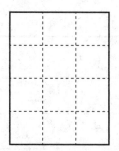

$\frac{1}{4} * \frac{2}{3} =$ _____

SRB 201

---

④ The table shows Chicago's typical monthly snowfall for several months.

Month	Snowfall (cm)
January	27.4
February	23.1
March	14.2
November	3.0
December	21.6

Of the months shown, which has the greatest typical snowfall?

_____

The least? _____

SRB 122–123

---

⑤ The graph below shows the number of garden tools needed for different numbers of groups participating in a gardening day. How many tools are needed if

4 groups participate? _____

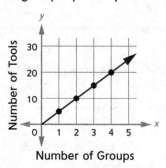

Number of Groups

SRB 55–56

---

⑥ $113.63 + 27.14 = ?$

_____
(estimate)

$113.63 + 27.14 =$ _____

SRB 128, 130

185

Fraction Multiplication Number Stories

Match each Number Sentence Card (*Math Masters*, page 199) with a Representation Card (*Math Masters*, page 200). Pick one pair and think of a number story that could be modeled by the number sentence and representation. Glue or tape the cards in the boxes below and record your number story and solution. Repeat with another pair of cards.

SRB
195–196

Number Sentence Card	Representation Card

Number Story: _____

_____

_____

_____

Number Sentence Card	Representation Card

Number Story: _____

_____

_____

Math Boxes

---

1 Solve. Show your work.

$3\frac{1}{4} + 7\frac{4}{5} =$ _____

SRB
177, 191

---

2 Jasmine's family is taking a bus to get to a baseball game that starts in 4 hours. The bus leaves in 1.5 hours and the trip takes 2.75 hours. Will Jasmine's family make it to the game on time?

Fill in the circle next to the best answer.

(A) They'll be a little early for the game.

(B) They'll be a little late for the game.

(C) They'll be on time for the game.

SRB
128

---

3 Carmen's mom is buying rice. Brand A's package contains $\frac{3}{4}$ lb. Brand B's package contains $\frac{2}{3}$ lb. Which brand has more rice?

Answer: _____

How much more? _____ pound

SRB
174, 176,
189–190

---

4 Solve.

a. $\frac{1}{5} * 25 =$ _____

b. $17 * \frac{3}{4} =$ _____

SRB
195–196,
199–200

---

5 **Writing/Reasoning** Write a number story that can be modeled by Problem 4b.

_____

_____

_____

SRB
178–180

Solving Fraction Division Problems

For Problems 1 and 2, write a number model using a letter for the unknown. Solve, showing your solution strategy with representations or drawings. Summarize your work with a division number model. Check your answer using multiplication and write a number sentence to show how you checked.

1. Two students equally share $\frac{1}{4}$ of a stick of clay. How much of the stick will each student get?

    Number model: _____

    Solution: Each student will get _____ of the clay stick.

    Summary number model: _____

    Check using multiplication: _____

2. Three families equally share $\frac{1}{3}$ of a community garden space. How much of the community garden space does each family use?

    Number model: _____

    Solution: Each family uses _____ of the community garden space.

    Summary number model: _____

    Check using multiplication: _____

3. When you divide a fraction by a whole number greater than 1, is the quotient larger or smaller than the fraction? Explain.

    _____

    _____

    _____

Try This

4. Write a number story for $\frac{1}{5} \div 2$. Solve your story. _____

    _____

    _____

Feeding a Blue Whale

Blue whales are the largest animals on Earth. A blue whale can grow to more than 30 meters in length and can weigh more than 180 metric tons. Despite its enormous size, a blue whale mainly eats tiny shrimp-like animals called krill.

This table shows the average amount of krill eaten by a blue whale during feeding season.

Write the data shown in the table as ordered pairs.
Plot the points on the grid and connect them with a line.

Days (x)	Total Krill Eaten (metric tons) (y)
2	7
4	14
6	21

Ordered pairs:

(_____ , _____)

(_____ , _____)

(_____ , _____)

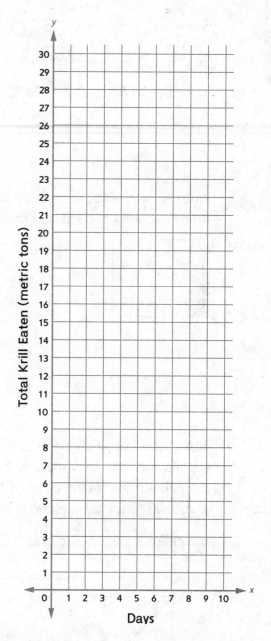

Use the graph to answer the questions.

1. About how many metric tons of krill does a blue whale eat in 1 day?

   _____ metric tons

2. About how many metric tons of krill does a blue whale eat in 1 week?

   _____ metric tons

3. About how many days does it take a blue whale to eat 28 metric tons of krill? _____ days

4. About how many days does it take a blue whale to eat 10 metric tons of krill? _____ days

Try This

About how many days does it take a blue whale to eat 7,000 kg of krill? Refer to the metric system chart in the *Student Reference Book* to help you.

_____

_____

189

Math Boxes

① Fill in the blanks with <, >, or = to make true number sentences.

a. $\frac{8}{10} * \frac{7}{8}$ _____ $\frac{8}{10}$

b. $\frac{6}{5} * 9$ _____ $1\frac{1}{5} * 9$

c. $1\frac{1}{12} * 76$ _____ 76

② Solve. Show your work.

$12\frac{1}{2} - 1\frac{3}{4} =$ _____

③ Write a multiplication number sentence that describes the shaded rectangle.

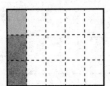

_____

④ The numbers below show the cost of a loaf of bread in different years between 1930 and 2008. Write the costs in order from least to greatest.

$0.09, $2.79, $0.70, $0.12, $0.25

_____, _____, _____,

_____, _____

⑤ The graph below shows how much flour Nigella has left after making batches of pancakes. How many cups of flour does she have left after making 6 batches of pancakes? _____ cups

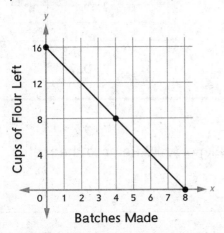

Batches Made

⑥ $54.19 - 36.57 = ?$

_____
(estimate)

$54.19 - 36.57 =$ _____

Fraction Division Problems

For Problems 1 and 2, write a number model using a letter for the unknown. Solve, showing your solution strategy. Summarize your work with a division number model. Check your answer using multiplication and write a number sentence to show how you checked.

1 How many $\frac{1}{2}$-pound boxes of nuts can be made from 10 pounds of nuts?

Number model: _____

Solution: _____ $\frac{1}{2}$-pound boxes of nuts can be made.

Summary number model: _____

Check with multiplication: _____

2 Darcy has 6 meters of yarn. She wants to cut the yarn into $\frac{1}{3}$-meter pieces to make necklaces with a kindergarten class. If she uses all 6 meters of yarn, how many $\frac{1}{3}$-meter pieces will Darcy have?

Number model: _____

Solution: Darcy will have _____ $\frac{1}{3}$-meter pieces of yarn.

Summary number model: _____

Check with multiplication: _____

191

Fraction Division Problems (continued)

3 Write a number story for $5 \div \frac{1}{4}$.

Number story: _____

_____

_____

Solve your number story. Draw a picture to show your solution strategy.

Solution: _____

Summary number model: _____

Check with multiplication: _____

4 When you divide a whole number by a fraction less than 1, is the quotient larger or smaller than the whole number? Explain.

_____

_____

_____

Math Boxes

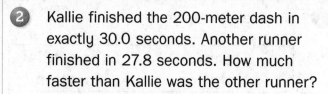

1 Solve. Show your work.

$$\underline{\hspace{2cm}} = 1\frac{7}{8} + 2\frac{1}{2}$$

SRB
177, 191

2 Kallie finished the 200-meter dash in exactly 30.0 seconds. Another runner finished in 27.8 seconds. How much faster than Kallie was the other runner?

_____
(estimate)

_____
(number model)

Answer: _____ seconds faster

SRB
44, 128,
131–132

3 Frances solved the problem $4\frac{7}{8} + 2\frac{1}{2}$ and got $6\frac{8}{10}$ as the sum. Is Frances correct? How do you know?

_____

_____

_____

SRB
181–185

4 The rectangle below is a model of Gary's garden. What is the area of his garden?

$\frac{5}{6}$ yd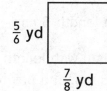

$\frac{7}{8}$ yd

_____
(number model)

Answer: _____ square yard

SRB
44, 202–
203, 225

5 **Writing/Reasoning** Explain how you solved Problem 1.

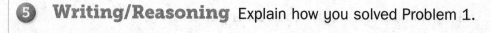

_____

_____

_____

_____

SRB
177, 191

Math Boxes

193

Math Boxes

① Complete the table.

Standard Notation	Exponential Notation
10,000	
	10^3
	10^8
1,000,000,000	
	10^5

SRB
68

② Solve.

$800 * 3,000 =$ _____

$54,000 =$ _____ $* 60$

$320,000 = 800 *$ _____

$5,000 *$ _____ $= 100,000$

$40 * 900 =$ _____

SRB
97–98

③ Which prism has the greatest volume? Choose the best answer.

◯ A cube with side length 5 cm.

◯ A rectangular prism with length of 5 cm, width of 5 cm, and height of 7 cm.

◯ A rectangular prism with a base area of 16 cm² and a height of 12 cm.

SRB
233

④ Fill in the blank with <, >, or =.

$8 \times \frac{3}{4}$ _____ 8

8 _____ 8×1.1

$8 \times \frac{9}{8}$ _____ 8

SRB
197–198

⑤ Place three fractions or mixed numbers on the number line.

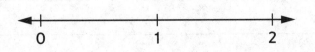

⑥ Look for a pattern.
Use it to solve the last problem below.

$300,000 \div 10 = 30,000$

$300,000 \div 100 = 3,000$

$300,000 \div 1,000 = 300$

$300,000 \div 10,000 =$ _____

SRB
158–160

SRB
95–96

Interpreting Dog Food Data

Samantha buys a 40-pound bag of dog food to feed her 3 dogs.
Her dogs eat a total of 7 pounds of dog food every 2 days.

1 Complete the table below. Write the data as ordered pairs.
Plot the points on the grid and connect them in a line.

Days (x)	Pounds of Dog Food Remaining (y)
0	40
2	
	26
6	

Ordered pairs:

(____, ____)

(____, ____)

(____, ____)

(____, ____)

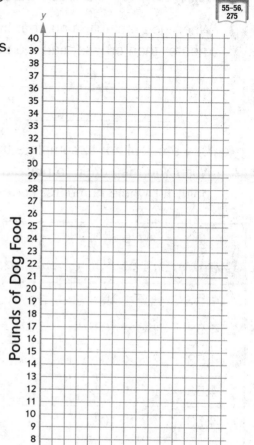

Use the grid to answer the questions.

2 About how many pounds of dog food will be left in the bag at the end of Day 5?

About _____ pounds

3 About how many pounds of dog food will be left in the bag at the end of Day 8?

About _____ pounds

4 Samantha's brother said that a 40-pound bag of dog food would last 2 weeks. Is he correct? Explain how you know.

_____

_____

Try This

Samantha's vet suggested that she make the 40-pound bag of food last for 2 weeks. Samantha thought, "I can plot a point at (14, 0) to show that there is no food left in the bag after 14 days." Plot a point at (14, 0). Draw a new line connecting just points (0, 40) and (14, 0). Use the new line to answer the questions.

5 On the new feeding plan, about how much food will be left in the bag after 2 days?

About _____ pounds

6 About how many pounds of food should Samantha feed her dogs every 2 days?

About _____ pounds

Multiplying and Dividing by Powers of 10

Multiplying by Powers of 10

SRB
133, 334

1. When you multiply a number by a power of 10, do you expect the product to be greater than or less than the start number? Why?

_____

_____

2. Use a calculator to complete the table. Look for patterns in how the decimal point moves. *Note:* You may need to place a zero in the tenths place to show the location of the decimal point for whole numbers. For example, write 453.0 instead of 453 to show the decimal point.

Start Number	× Power of 10	Result in Standard Notation	Movement of Decimal Point	
			Direction	*Number of Places*
4.53	× 10^1			
4.53	× 10^2			
4.53	× 10^3			
4.53	× 10^4			
4.53	× 10^5			
4.53	× 10^6			

3. **a.** Look at your results in the table above. Compare the power of 10 in each row to the movement of the decimal point. What do you notice?

_____

_____

b. Use the patterns you noticed to write a rule for multiplying any decimal by a power of 10.

_____

_____

_____

4. If you *divided* a start number by a power of 10, would you expect the quotient to be greater than or less than the start number? Why?

_____

_____

196

Multiplying and Dividing by Powers of 10 (continued)

Dividing by Powers of 10

5 Use a calculator to complete the table. Look for patterns in how the decimal point moves.

Start Number	÷ Power of 10	Result in Standard Notation	Movement of Decimal Point	
			Direction	*Number of Places*
67.2	$\div 10^1$			
67.2	$\div 10^2$			
67.2	$\div 10^3$			
67.2	$\div 10^4$			
67.2	$\div 10^5$			
67.2	$\div 10^6$			

6 **a.** Look at your results in the table above. Compare the power of 10 in each row to the movement of the decimal point. What do you notice?

_____

_____

b. Use the patterns you noticed to write a rule for dividing any decimal by a power of 10.

_____

_____

_____

Applying Rules for Multiplying and Dividing by Powers of 10

Use the rules you discovered to multiply and divide in Problems 7–12. Do not use a calculator.

7 $5.8 \times 10^2 =$ _____

8 $2.8 \div 10^2 =$ _____

9 $673.9 \div 10^2 =$ _____

10 $23.7 \times 10^2 =$ _____

11 $3.1 \times 10^4 =$ _____

12 $49.2 \div 10^4 =$ _____

13 Explain the placement of the decimal point in your answer for Problem 7.

_____

_____

_____

Math Boxes

1   Solve.

a. $\frac{2}{3}$ of 7 = _____

b. $\frac{3}{8}$ of 5 = _____

c. $\frac{4}{5}$ of 12 = _____

SRB
196

2   **a.** Write a 6-digit number with 7 in the thousands place, 5 in the hundredths place, 4 in the tenths place, 3 in the tens place, and 9s in all other places.

_____

b. Write this number in words.

_____

_____

_____

SRB
117–119

3   A costume designer is using exactly 264 yards of green fabric to make 36 frog costumes. If each costume requires the same amount of fabric, how many yards will be used for each one?

_____
(number model)

Answer: _____
SRB
44, 109,
113–114

4   Nigel has 2 dogs. One eats $2\frac{1}{2}$ pounds of food each week. The other eats $1\frac{3}{8}$ pounds each week. Together, how much food do the dogs eat in a week?

_____
(number model)

Answer: _____ pounds
SRB
177–180,
191

5   **Writing/Reasoning** Explain how you decided what to do with the remainder in Problem 3.

_____

_____

_____

_____

SRB
113–114

Practicing Fraction Division

For each problem, write a number model using a letter for the unknown. Solve and show your solution strategy with a drawing or other model. Check your answer using multiplication, and write a number sentence to show how you checked.

SRB
207–209

1 Selena has 6 inches of tape to attach photos to her display board. If she cuts the tape into $\frac{1}{2}$-inch pieces, how many pieces will she have?

Number model: _____

Selena will have _____ $\frac{1}{2}$-inch pieces of tape.

Check: _____

2 Heather has $\frac{1}{4}$ of a watermelon. If she shares the watermelon equally with her brother and sister, how much watermelon will each person get?

Number model: _____

Each person will get _____ watermelon.

Check: _____

3 Mr. Middleton has $\frac{1}{2}$ box of colored paper. He wants to split the paper equally among 5 table groups for an art project. How much paper will each table group get?

Number model: _____

Each group will get _____ box of paper.

Check: _____

4 One box of raisins contains 2 cups. A serving is $\frac{1}{4}$ cup. How many servings are in one box?

Number model: _____

There are _____ servings in one box.

Check: _____

199

Math Boxes

1 Priya wants to use one-third of her bead collection to make bracelets for 5 friends. What fraction of her collection should she use for each bracelet?

_____
(number model)

Answer: _____

SRB
207

2 Solve.

a.   5.7 1
    + 9.8 8
    _____

b.   8 0 3.4
    +   9 8.6
    _____

SRB
130

3 Write 3 different common denominators you could use to solve $1\frac{2}{3} + \frac{5}{6}$.

_____

SRB
177

4 Round each decimal to the nearest tenth.

a. 4.75 _____

b. 17.31 _____

c. 9.18 _____

d. 0.06 _____

SRB
124–127

5 Delaney is keeping track of the total number of vegetable servings she has eaten for the week. Plot the data on the grid and answer the question.

Day (x)	Total Vegetable Servings (y)
1	2
2	4
3	6
4	8
5	10
6	12
7	14

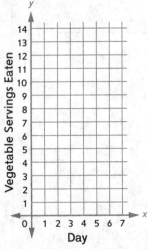

On what day did Delaney eat her seventh serving of vegetables? _____

SRB
55–56,
275

200

Converting Metric Units

Math Message

1 There are 100 centimeters (cm) in a
meter (m). Use this information to write
a rule in the function machine. Then
complete the "What's My Rule?" table
at the right.

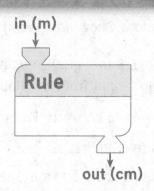

in (m)

Rule

out (cm)

in (m)	out (cm)
1	100
2	
3	
4	
5	

SRB
215–216,
328

Convert between the given units to complete the "What's My Rule?" tables below.
Use exponential notation to write each rule.

2 **a.** Convert between centimeters (cm) and
millimeters (mm).

in

Rule

out

in (cm)	out (mm)
1	10
3	
4.7	
	50
	5
	17

b. Write a rule you could use to convert
from millimeters to centimeters.

Hint: How can you find the *in* number
if you know the *out* number?

_____

3 **a.** Convert between grams (g) and
kilograms (kg).

in

Rule

out

in (g)	out (kg)
1,000	1
2,500	
250	
	8
	0.8
	0.08

b. Write a rule you could use to convert
from kilograms to grams.

Hint: How can you find the *in* number
if you know the *out* number?

_____

4 There are 43 milligrams of caffeine in a bottle of iced tea. How many grams of caffeine is
that? Answer the questions below to solve.

in

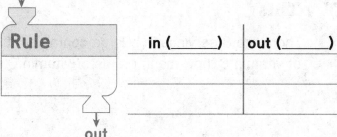

Rule

out

a. What units do you need to convert?

From _____ to _____

in (____)	out (____)

b. How are those units related?

_____ = _____

c. Write your answers to Parts a and b in the "What's My Rule?" table. Fill in the rule.

d. How many grams is 43 milligrams? _____

201

Solving Conversion Stories

Solve the number stories below. Show your work. Label the units for each step.

SRB
133, 136,
215–216

① Max has a shelf that is 1.3 m high. If he places a lamp that is 45 cm high on top of the shelf, what is the height from the floor to the top of the lamp in meters?

- What units do you need to convert? From _____ to _____

- How are those units related? _____ = _____

- What rule can you use to convert from the units you have to the units you need to

  solve the problem? _____

The height from the floor to the top of the lamp is _____ m.

② Alida has a cough. Her mother gave her 10 mL of cough syrup and 0.5 L of water. How many mL of liquid was Alida given in all?

③ Tina and Kyle are doing a 5 km walk for charity. They passed a sign that said, "Only 100 meters to go!" How many km had Tina and Kyle walked when they passed the sign?

Alida was given _____ mL of liquid.

Tina and Kyle had walked _____ km when they passed the sign.

Try This

④ There are 8 servings in a large container of orange juice. If each serving has 60 mg of vitamin C, how many grams of vitamin C are in the whole container?

There are _____ g of vitamin C in the container.

Math Boxes

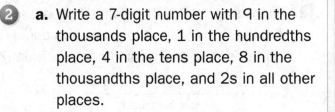

Math Boxes

1 Which number sentence matches the picture below?

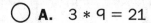

Fill in the circle next to the best answer.

○ **A.** $3 * 9 = 21$

○ **B.** $3 * 7 = \frac{21}{9}$

○ **C.** $3 * \frac{7}{9} = \frac{21}{9}$

SRB
199

2 **a.** Write a 7-digit number with 9 in the thousands place, 1 in the hundredths place, 4 in the tens place, 8 in the thousandths place, and 2s in all other places.

_____

b. Write this number in words.

_____

_____

_____

SRB
117–119

3 Juan is driving 732 miles. If Juan's car travels about 31 miles on each gallon of gasoline, about how many gallons of gas will it take him to make the entire trip?

_____
(number model)

Answer: about _____ gallons

SRB
44, 109.
113–114

4 A puppy weighed $\frac{7}{8}$ lb at birth and $2\frac{1}{4}$ lb at 2 weeks of age. How much weight did she gain from birth to the age of 2 weeks?

_____
(number model)

Answer: _____ lb

SRB
177–180,
192–193

5 **Writing/Reasoning** How would the value of the 4 in Problem 2 change if it were moved one place to the left? If it were moved one place to the right? Explain.

_____

_____

_____

_____

SRB
118–119

203

Math Boxes

① Quincy has $\frac{1}{2}$ box of cereal to eat over 6 days. If he splits the $\frac{1}{2}$ box into equal portions, how much will he eat each day?

_____

(number model)

SRB
207

Answer: _____ box of cereal

② Solve.

a.
```
  6 0 4 . 2 4
− 5 9 9 . 7 9
─────────────
```

b.
```
          9
−   3 . 5 2
─────────────
```

SRB
131–132

③ Rewrite the problem with a common denominator. Then solve.

$12\frac{1}{2} - 9\frac{3}{7} = ?$

_____

$12\frac{1}{2} - 9\frac{3}{7} =$ _____

SRB
177,
192–193

④ Round 2,813.965 to:

a. the nearest hundred _____

b. the nearest hundredth _____

c. the nearest one _____

d. the nearest tenth _____

SRB
79–82,
124–127

⑤ Use the points on the grid to fill in the missing coordinates in the table.

SRB
55–56,
275

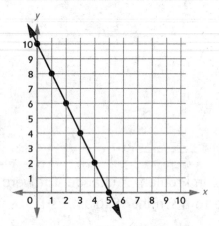

x	y
0	
	8
2	6
3	
	2
5	

Write a number story that could be represented by this data.

_____

_____

_____

Line Plot: Pencil-Length Data

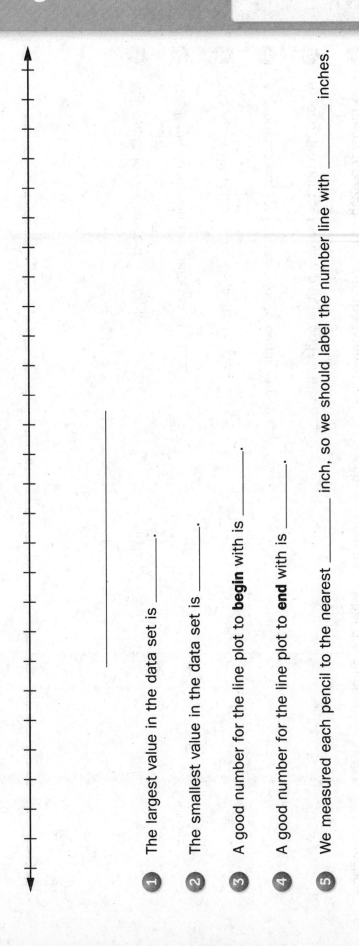

1. The largest value in the data set is _____.

2. The smallest value in the data set is _____.

3. A good number for the line plot to **begin** with is _____.

4. A good number for the line plot to **end** with is _____.

5. We measured each pencil to the nearest _____ inch, so we should label the number line with _____ inches.

SRB
244-245

1   My height to the nearest $\frac{1}{2}$ inch is _____.

2   The largest value in the data set is _____.

3   The smallest value in the data set is _____.

4   A good number for the line plot to **begin** with is _____.

5   A good number for the line plot to **end** with is _____.

6   We measured each student to the nearest _____ inch, so we should label the number line with _____ inches.

206

Using a Line Plot to Solve Problems

Use your line plot about student height data to solve the following problems.

SRB
187–188,
247, 328

1 How many students are included in your data set? _____

2 What is the most common height represented in your data set? _____ inches

3 How many students in your class are 55 or more inches tall? _____

4 What is the difference between the largest height measurement and the smallest height

measurement? _____ inches

Explain how you solved the problem.

_____

_____

5 What is the total combined height of all the students in your class? _____ inches

Show how you solved the problem.

6 Imagine that you and your classmates decided to lie down on a baseball field, starting at
home plate and placing yourselves head-to-toe along the first base line. How close would
you come to reaching first base? (The distance from home plate to first base is 90 feet.)

☐ We would cover the exact distance from home plate to first base.

☐ We would not reach all the way to first base. We would be short by _____.

☐ We would go past first base. We would go past by _____.

Explain how you solved the problem.

_____

_____

Animal-Height Line Plots

Emperor Penguin Heights

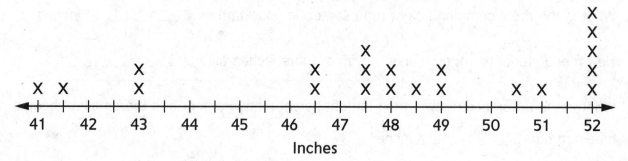

Inches

Chimpanzee Heights

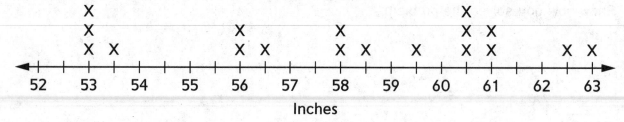

Inches

Red Kangaroo Heights

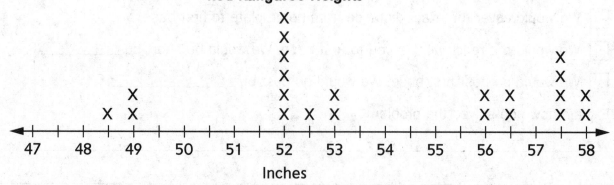

Inches

Using a Line Plot to Even Out Data

Use the data in the line plot on journal page 208 to find the typical height of a(n)

SRB
109, 187, 247-248

_____.
(name of animal)

Answer these questions to help you find the animal's typical height. Show your work in the space below.

1 What is the combined total of all the measurements shown on your animal's line plot?

_____ inches

2 How many data points are shown on the line plot? _____ data points

3 Write a number model that shows how to even out the data to find the typical height of your animal. (Use your answers to Problems 1 and 2.)

_____

4 The typical height of a(n) _____ is about _____ inches.
(name of animal)

5 Fill in the name of your animal and circle the word that completes the sentence based on what you found out.

A typical _____ is **taller shorter** than a typical fifth grader.
(name of animal)   (circle one)

Practicing Fraction Multiplication

For Problems 1–4, write a number model using a letter for the unknown. Then solve.

SRB
199,
202–203

① Hannah bought a package of 12 pens. She took $\frac{2}{3}$ of the pens to school. How many pens did Hannah take to school?

Number model: _____

② Craig is walking to a grocery store that is $\frac{3}{4}$ mile away. When he was $\frac{1}{2}$ of the way to the store, he ran into his friend Greg. How far had Craig walked when he saw Greg?

Number model: _____

Hannah took _____ pens to school.

Craig had walked _____ mile.

③ Lamar is researching postage stamps from different years. One stamp measured $\frac{7}{8}$ inch by $\frac{2}{3}$ inch. What is the area of the stamp?

Number model: _____

④ Phoebe's class has filled $\frac{3}{4}$ of a box with food donations. The class wants to collect 4 times as much by the end of the food drive. How many boxes does Phoebe's class want to fill?

Number model: _____

The area of the stamp is _____ in².

Phoebe's class wants to fill _____ boxes.

⑤ Write a number story that could be modeled by this number sentence. Then solve.

$$\frac{5}{6} * \frac{1}{2} = p$$

_____

_____

_____

_____

Solution: _____

Math Boxes

① Shade a rectangle to represent $\frac{2}{3} * \frac{5}{6}$.

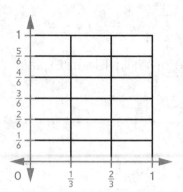

$\frac{2}{3} * \frac{5}{6} =$ _____

SRB
202

② Solve.

$6 \div \frac{1}{3} =$ _____

SRB
209

③ Write the following numbers in order from greatest to least.

0.38   0.308   3.08   3.38   0.038

_____ _____ _____

_____ _____

SRB
121–123

④ Fill in the blank with <, >, or =.

a. $15 * \frac{2}{3}$ _____ 15

b. $15 * \frac{12}{6}$ _____ 15

c. $15 * \frac{18}{18}$ _____ 15

SRB
197–198

⑤ **Writing/Reasoning** Explain how you solved Problem 4c.

_____

_____

_____

_____

SRB
197–198

211

Finding the Volume
of Willis Tower

1 What is the approximate volume of Willis Tower?

About _____ cubic feet

SRB
18-21,
233-234

2 Describe the strategy your group used to estimate the volume of Willis Tower. Explain
your strategy as clearly as you can.

_____

_____

_____

_____

_____

_____

_____

_____

_____

_____

_____

3 Do you think your group could have used a more efficient strategy? Explain at least one way
your strategy could have been more efficient.

_____

_____

_____

_____

_____

_____

_____

Math Boxes

1 Emmy bought a book on sale for $12.50. The regular price of the book was $16.99. How much money did she save?

SRB
131–132

Answer: $_____

2 What is $\frac{11}{15}$ of 75?

SRB
196

Answer: _____

3 Orly had $\frac{3}{4}$ gallon of paint. She spilled half the paint on the floor. How much paint did she spill?

_____
(number model)

SRB
195,
201–203

Answer: _____ gallon

4 Solve.

a. $\frac{5}{6} - \frac{2}{3} =$ _____

b. $1\frac{3}{8} - \frac{3}{4} =$ _____

SRB
177, 189–
190, 192

5 Write 471.32 in expanded form, and then in words.

Expanded form:

_____

_____

Words:

_____

_____

SRB
117–118

6 Solve.

a. $8.4 * 10^1 =$ _____

b. $0.84 * 10^5 =$ _____

c. $84 * 10^2 =$ _____

d. $8.4 * 10^3 =$ _____

SRB
133

Math Boxes

213

Calibrating a Bottle

SRB
236, 238

Materials

☐ 2-liter bottle with the top cut off

☐ bucket or bowl with about 2 liters of water

☐ measuring cup marked in milliliters

☐ ruler   ☐ scissors

☐ paper   ☐ tape

1. Fill the bottle with about 5 inches of water.
   (That's a little more than halfway.)

2. Cut a 1-inch by 6-inch strip of paper. Tape the strip
   to the outside of the bottle with one end at the top of
   the bottle. The other end should be below the water level.

   Add water
   100 mL
   at a time.

   700 mL
   600 mL
   500 mL
   400 mL
   300 mL
   200 mL
   100 mL
   0 mL

   9"

   5"

3. Mark the paper strip at the water level. Write "0 mL"
   next to the mark.

4. Use the measuring cup to pour exactly 100 mL of
   water into the bottle. Mark the paper strip at the new
   water level. Write "100 mL" next to the mark.

5. Pour another 100 mL of water into the bottle. Mark the new water level and label the mark
   with "200 mL."

6. Continue adding 100 mL of water at a time until the water is less than 1 inch from the top of
   the bottle. Mark and label each new water level.

7. Pour water out of the bottle until the water level is back at the 0 mL mark.

The calibrated bottle you have created is a tool for measuring volume.

How do you think this tool could be used to find the volume of the orange from the Math
Message? *Hint:* What would happen if you put the orange in the bottle?

_____

_____

_____

Record the volume of the orange. _____ mL

Measuring Volume by Displacement

1. Use the displacement method to find the volume in milliliters of 1 base-10 flat, 2 base-10 flats, and 3 base-10 flats. Record the volumes in the Volume (mL) column of the table below.

2. Find the volume in cubic centimeters (cm^3) of 1 base-10 flat, 2 base-10 flats, and 3 base-10 flats. Record the volumes in the Volume (cm^3) column of the table below.

   Remember: The volume of one centimeter cube is 1 cm^3.

Number of Base-10 Flats	Volume (mL)	Volume (cm^3)
1		
2		
3		

3. Look at the table. What do you notice?

   _____

4. Use your observation in Problem 3 to fill in the blank: 1 mL = _____ cm^3

5. Use the calibrated bottle to measure the volume of the 3 larger objects at your workstation. Record the volumes in milliliters and cubic centimeters in the table below.

Object	Volume (mL)	Volume (cm^3)

6. At your workstation are several small, identical objects. Use the displacement method to find the volume of one of these objects. What do you notice?

   _____

   _____

7. Does displacing 1 object give you a precise measurement? Explain. _____

   _____

8. Try this strategy for finding a more precise measure of the volume.

   a. Place all of the identical objects in the bottle. What is the volume of all the objects?

      _____ mL, or _____ cm^3

   b. How many of the objects did you put in the bottle? _____

   c. Divide your answer in Part a by your answer in Part b to estimate the volume of just one object. _____ mL, or _____ cm^3

215

Math Boxes

① Shade a rectangle to represent $\frac{7}{8} * \frac{3}{4}$.

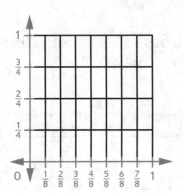

What is $\frac{7}{8} * \frac{3}{4}$? _____

SRB
202

② Solve.

$3 \div \frac{1}{8} =$ _____

SRB
209

③ Fill in the circle(s) next to the true number sentence(s).

○ **A.** 7.4 > 0.74 > 0.074

○ **B.** 1.23 < 123 < 12.3

○ **C.** 0.518 < 5.18 < 51.8

○ **D.** 91.52 > 915.2 > 9,152

SRB
121–123

④ Fill in the blank with >, <, or =.

a. $\frac{5}{7}$ _____ $\frac{5}{7} * \frac{6}{6}$

b. $\frac{5}{7}$ _____ $\frac{5}{7} * \frac{4}{3}$

c. $\frac{5}{7}$ _____ $\frac{5}{7} * \frac{6}{7}$

SRB
197–198

⑤ **Writing/Reasoning** Write a number story that could be modeled by the number sentence in Problem 2.

_____

_____

_____

_____

_____

SRB
209–210

Math Boxes

Estimating Decimal Products and Quotients

The digits provided in the Answer column are correct, but they are missing a decimal point. For each problem, write a number sentence to estimate the product or quotient. Use your estimate to place a decimal point in the digits provided. An example is done for you.

SRB
128

Problem	Estimation Number Sentence	Answer (place the decimal point)
Example: 12.2 * 1.9	$10 * 2 = 20$	2 3.1 8
① 17.4 * 97.5		1 6 9 6 5
② 83.12 * 7.25		6 0 2 6 2
③ 0.36 * 325.5		1 1 7 1 8
④ 4.85 * 0.6		2 9 1
⑤ 1.8 * 27.3		4 9 1 4
⑥ 95.76 ÷ 7.6		1 2 6
⑦ 515.87 ÷ 65.3		7 9
⑧ 2.76 ÷ 3.68		0 7 5
⑨ 101.8 ÷ 0.8		1 2 7 2 5
⑩ 1,390.72 ÷ 21.73		6 4 0 0

Try This

⑪ Four families held a garage sale and split the proceeds evenly. They made a total of $1,256.60 at the garage sale. Circle the number model that represents the correct amount of money each family earned.

     a. $1,256.60 ÷ 4 = $31.41   **b.** $1,256.60 ÷ 4 = $314.15

     c. $1,256.60 ÷ 4 = $3,141.50   **d.** $1,256.60 ÷ 4 = $3.14

⑫ How did you use estimation to think about Problem 11?

_____

_____

Math Boxes

---

① The average height of men in the United States is 69.3 in. The average height of women in the United States is 63.8 in. How much taller is the average man than the average woman?

_____
(number model)

SRB
44,
131–132

Answer: _____ inches

---

② What is $\frac{11}{12}$ of 6?

SRB
196

Answer: _____

---

③ A library spent $\frac{5}{8}$ of its budget to buy new books. They set aside $\frac{1}{12}$ of the book money to buy young adult novels. What part of the budget went towards buying young adult novels?

_____
(number model)

Answer: _____ of the budget

SRB
195,
201–203

---

④ Solve.

a. $3\frac{1}{4} - 1\frac{5}{6} =$ _____

b. $\frac{6}{7} - \frac{5}{9} =$ _____

SRB
177, 189–
190, 192

---

⑤ a. Write the decimal 1,072.039 in expanded form.

_____

_____

_____

b. Write the decimal 1,072.039 in words.

_____

_____

SRB
117–118

---

⑥ Solve.

a. $0.025 * 10^3 =$ _____

b. $10^5 * 0.25 =$ _____

c. $2.5 * 10^2 =$ _____

d. $0.025 * 10^1 =$ _____

SRB
133

Math Boxes

Multiplying Decimals: Estimation Strategy

Math Message

SRB
100, 102,
128, 134

Tori has 8 blocks. Each block is 1.2 centimeters high.

If Tori stacks the blocks, what will the height of the stack be? _____ cm

Solve Problems 1–3 using the following method:

Step 1: Make an estimate.

Step 2: Multiply as if the factors were whole numbers.

Step 3: Use your estimate to place the decimal point in the product.

1 76.1 * 9.6 = ?

   Estimate: _____

   Answer: _____

2 189.6 * 1.75 = ?

   Estimate: _____

   Answer: _____

3 5.6 * 0.8 = ?

   Estimate: _____

   Answer: _____

Multiplying Decimals: Shifting the Decimal Point

SRB
100, 102,
133, 135

Solve Problems 1–3 using the following method:

Step 1: Multiply both factors by a power of 10 to make them whole numbers.

Step 2: Multiply the whole numbers.

Step 3: Undo the multiplication by powers of 10 by dividing the product by the
same powers of 10.

1   76.1 * 9.6 = ?

   a. 76.1 * 10$^{\square}$ = _____   9.6 * 10$^{\square}$ = _____

   b. Answer to whole-number problem: _____

   c. Answer to decimal problem: 76.1 * 9.6 = _____

---

2   189.6 * 1.75 = ?

   a. 189.6 * 10$^{\square}$ = _____   1.75 * 10$^{\square}$ = _____

   b. Answer to whole-number problem: _____

   c. Answer to decimal problem: 189.6 * 1.75 = _____

---

3   5.6 * 0.8 = ?

   a. 5.6 * 10$^{\square}$ = _____   0.8 * 10$^{\square}$ = _____

   b. Answer to the whole-number problem: _____

   c. Answer to decimal problem: 5.6 * 0.8 = _____

Math Boxes

Math Boxes

1   Estimate the quotient for each problem. Then circle the most reasonable answer.

     a. $83.7 \div 3 = ?$

        2.79   27.9   279.0

     b. $13.56 \div 0.8 = ?$

        1.695   16.95   169.5

SRB
128

2   Solve. Use the area model to help you.

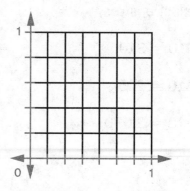

$\dfrac{3}{5} * \dfrac{5}{7} =$ _____

SRB
202

3   A green anaconda can grow up to 8.8 meters in length. A coral snake can grow up to 0.76 meter in length. How much longer can a green anaconda grow than a coral snake?

_____
          (number model)

Answer: _____ meters longer

SRB
44,
131–132

4   Solve.

$\dfrac{1}{4} \div 4 = ?$

Answer: _____

SRB
207

5   **Writing/Reasoning** Write a number story that can be modeled by Problem 4.

_____

_____

_____

_____

SRB
207

Which Answer Makes Sense?

For each set of problems, circle the number sentence with the correct product. Do not calculate the exact answer. Explain how you knew which product was correct without finding the exact answer.

 1   0.5 * 410 = 810

 0.5 * 410 = 205

 0.5 * 410 = 2,050

2   1 * 410 = 410

 1 * 410 = 4,130

 1 * 410 = 41

3   2.5 * 410 = 750

 2.5 * 410 = 330

 2.5 * 410 = 1,025

222

Math Boxes: Preview for Unit 7

Math Boxes

---

① Find the area of the rectangle below.

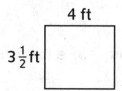

4 ft

$3\frac{1}{2}$ ft

Area = _____ ft²

SRB
225

---

② Draw lines to match the mixed numbers and equivalent fractions.

Mixed Number	Fraction
$5\frac{3}{8}$	$\frac{20}{8}$
$1\frac{7}{8}$	$\frac{15}{8}$
$2\frac{4}{8}$	$\frac{43}{8}$
$7\frac{1}{8}$	$\frac{55}{8}$
$6\frac{7}{8}$	$\frac{57}{8}$

SRB
171–172

---

③ The image below is a trapezoid. Name 3 attributes of the trapezoid.

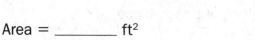

_____

_____

_____

SRB
268

---

④ Fill in the blanks for each pattern.

a. 5, 14, 23, _____,

_____, _____

b. 12, 72, _____, 192,

_____, _____

c. _____, 21.50, 23.00,

_____, _____

SRB
51

---

⑤ Complete the "What's My Rule?" table and state the rule.

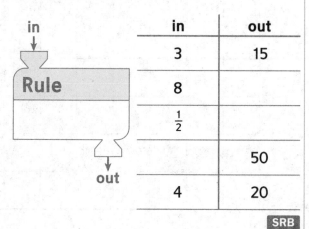

in	out
3	15
8	
$\frac{1}{2}$	
	50
4	20

SRB
53–54

---

⑥ Rewrite each whole number as a fraction with the given denominator.

a. $3 = \dfrac{\boxed{}}{6}$

b. $5 = \dfrac{\boxed{}}{4}$

c. $12 = \dfrac{\boxed{}}{5}$

SRB
171

223

Dividing Decimals

For Problems 1 and 2:

- Write a number model.

- Make an estimate. Write a number sentence to record your estimate.

- Divide as if the dividend were a whole number. If there is a remainder, write it as a fraction and use it to round the quotient to the nearest whole number.

- Use your estimate to place the decimal point. Record your answer.

1 Three sisters set up a lemonade stand. On Wednesday they made $8.46. If they share the money equally, how much will each sister get?

Number model: _____

Estimate: _____

2 Janine is building a bookshelf. She has a board that is 6.77 meters long. She wants to cut it into 5 pieces of equal length. How long will each piece be?

Number model: _____

Estimate: _____

Answer: Each sister will get _____.

Answer: Each piece is about _____ meters long.

224

For Problems 3–6:

- Make an estimate. Write a number sentence to record your estimate.

- Divide as if the dividend were a whole number. Show your work on the computation grid. If there is a remainder, write it as a fraction and use it to round the quotient to the nearest whole number.

- Use your estimate to place the decimal point. Record your answer.

SRB
128,
137–139

3 9.44 / 4 = ?

Estimate: _____

9.44 / 4 = _____

4 46.8 ÷ 12 = ?

Estimate: _____

46.8 ÷ 12 = _____

5 89.9 ÷ 4 = ?

Estimate: _____

89.9 ÷ 4 ≈ _____

6 253.7 / 6 = ?

Estimate: _____

253.7 / 6 ≈ _____

Math Boxes

Math Boxes

(1) Antonio was practicing the standing long jump for gym. He jumped 8 times and wrote down the following measurements in inches: $37\frac{1}{2}$, $36\frac{1}{2}$, 37, $36\frac{1}{2}$, 38, $38\frac{1}{2}$, $36\frac{1}{2}$, 37. Use the data to complete the line plot.

What is the difference between Antonio's longest jump and his shortest jump?

_____ inches

Antonio's Long Jumps

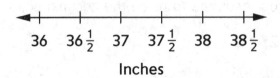

36    $36\frac{1}{2}$ 37 $37\frac{1}{2}$ 38 $38\frac{1}{2}$

Inches

SRB
188 244–
245, 247

(2) Fill in the missing exponent to make a true number sentence.

a. $7.2 \times 10^{\square} = 72{,}000$

b. $15.3 \times 10^{\square} = 1{,}530$

c. $0.84 \times 10^{\square} = 8.4$

SRB
133

(3) $4\frac{2}{3} + 5\frac{3}{5} = ?$

Circle ALL correct answers.

A. $9\frac{5}{8}$ **B.** $9\frac{19}{15}$

C. $9\frac{4}{15}$ **D.** $10\frac{4}{15}$

SRB
191

(4) In the table below, Clara is recording the total distance she has run at track practice. Plot the data on the grid, connect the points with a line, and answer the question.

Day (x)	Total Miles (y)
1	3
2	6
3	9
4	12

If the patterns in the table continue, on what day will Clara run her 18th mile?

_____

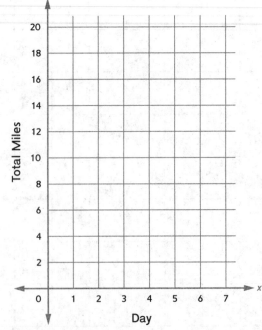

SRB
55–56,
275

Dividing by Decimals

For Problems 1 and 2:

SRB
128,
140–141

- Write an equivalent division problem that has a whole-number divisor. Be sure to multiply the dividend and the divisor by the same number.

- Make an estimate for your equivalent problem.

- Divide as if the dividend were a whole number to solve your equivalent problem.

- Use your estimate to place the decimal point.

- Complete the number sentences to show the answers to the equivalent problem and the original problem.

 1 $2.79 \div 0.9 = ?$

Equivalent problem:

_____

Estimate:

_____

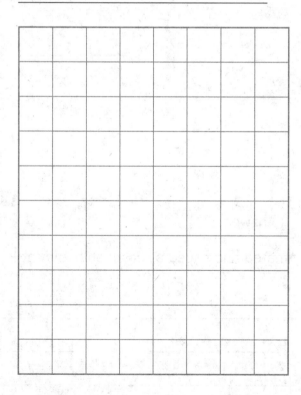

Equivalent problem with answer:

_____ ÷ _____ = _____

$2.79 \div 0.9 =$ _____

 2 $85.4 \div 0.14 = ?$

Equivalent problem:

_____

Estimate:

_____

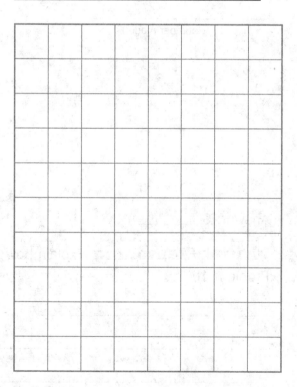

Equivalent problem with answer:

_____ ÷ _____ = _____

$85.4 \div 0.14 =$ _____

Math Boxes

1 Estimate the quotient for each problem. Then circle the most reasonable answer.

a. $30.6 \div 9 = ?$

   3.4   34.0   0.340

b. $506.9 * 1.2 = ?$

   60.828   608.28   6,082.8

c. $954.16 \div 238.54 = ?$

   40   4   0.4

SRB
128,
138–139

2 Solve. Use the area model to help you.

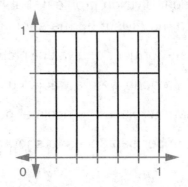

$\frac{2}{3} * \frac{5}{6} = $ _____

SRB
202

3 Eloise bought $79.84 worth of groceries. What was her change if she paid with a $100 bill?

_____
(number model)

Answer: $ _____

SRB
44,
131–132

4 Solve.

$\frac{1}{3} \div 9 = ?$

Answer: _____

SRB
207

5 **Writing/Reasoning** Explain how you estimated and chose a reasonable answer in Problem 1b.

_____

_____

_____

_____

SRB
128,
138–139

Math Boxes

Math Boxes

Math Boxes

1 The data in the table show how many women's shoes of each size were sold at a shoe store in one day. Use the data to make a line plot. Then answer the questions.

Shoes Sold

Sizes Sold	
5	///
$5\frac{1}{2}$	//
6	~HH
$6\frac{1}{2}$	
7	~HH /
$7\frac{1}{2}$	~HH //
8	~HH //
$8\frac{1}{2}$	////
9	//

```
  ◄─┼────┼────┼────┼────┼────┼────┼────┼────►
    5   5½   6   6½   7   7½   8   8½   9
                    Shoe Sizes
```

Which size shoe was not sold on this day? _____

Which size(s) were most frequently sold on this day? _____

SRB
244–
245, 247

---

2 Fill in the exponent to make a true number sentence.

a. $3{,}700 \div 10^{\square} = 37$

b. $67{,}536 \div 10^{\square} = 6.7536$

c. $712 \div 10^{\square} = 0.712$

SRB
136

3 $9\frac{3}{5} - 8\frac{6}{7} = ?$

Choose the best answer.

◯ $\frac{26}{35}$ ◯ $1\frac{9}{35}$

◯ $1\frac{26}{35}$ ◯ $\frac{9}{35}$

SRB
192–193

---

4 Darian has a $20 allowance to pay for bus fare. He is keeping track of his spending in the table. Plot the data from the table on the grid, connect the points with a line, and answer the question.

Number of Bus Rides	Money Left ($)
0	20
1	18
2	16
3	14

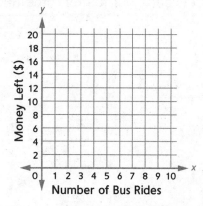

Number of Bus Rides

How many bus rides will Darian be able to take

with $20? _____

SRB
55–56,
275

229

Estimating Your Reaction Time

SRB
121, 129,
131–132

It takes two people to perform this experiment. The Tester holds the Grab-It Gauge at the top. The Contestant gets ready to catch the gauge by placing his or her thumb and index finger at the bottom of the gauge *without quite touching it*. (See *diagram*.)

When the Contestant is ready, the Tester lets go of the gauge. The Contestant grabs it as quickly as possible with his or her thumb and index finger.

The number grabbed by the Contestant shows that person's reaction time to the nearest hundredth of a second. The Contestant then records that reaction time in the data table shown below.

Partners take turns being Tester and Contestant. Each person should perform the experiment 10 times using his or her right hand.

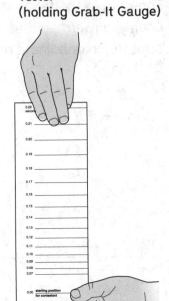

Tester
(holding Grab-It Gauge)

Contestant
(not quite touching
Grab-It Gauge)

Right-Hand Reaction Times (in seconds)				
1.	2.	3.	4.	5.
6.	7.	8.	9.	10.

1 Write your reaction times in order from fastest to slowest.

_____ _____ _____ _____ _____ _____ _____ _____ _____ _____

(fastest)   (slowest)

2 What is the difference between your fastest time and your slowest time? _____ sec

Use the results of your Grab-It experiment to answer the following questions.
Remember: "sec" is the abbreviation for seconds.

SRB
130, 137,
247-248

3 Which time came up most often in your results? _____ sec

4 Look at your times in order from fastest to slowest. What time is in the middle?

Hint: Your middle time might fall between two of the times you have listed.

_____

5 Your class decided on a strategy to identify each person's typical reaction time.

a. Which strategy did your class choose? _____

b. Why did your class choose this strategy?

_____

_____

c. Find your typical right-hand reaction time using the strategy chosen. Show your work.

Your typical reaction time: _____ sec

6 How long will it take for a hand squeeze to travel around your whole class? Use your class's Typical Reaction Time line-plot data to help you predict your class's total reaction time. Write an expression to model the prediction, using grouping symbols if necessary. Evaluate your expression to solve.

Expression: _____

Estimated class reaction time: _____ sec

Math Boxes

---

1 What is the area of the rectangle?

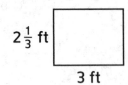

$2\frac{1}{3}$ ft

3 ft

SRB
225

Area: _____ ft²

---

2 Write each fraction as a mixed number.

a. $\frac{53}{12} =$ _____

b. $\frac{64}{7} =$ _____

c. $\frac{47}{6} =$ _____

SRB
171–172

---

3 a. Name one attribute shared by a square and a rhombus.

_____

_____

b. Name one attribute of a square that is not an attribute of a rhombus.

_____

_____

SRB
268

---

4 Write the next two numbers in each pattern.

a. 112, 56, 28, _____, _____

b. $\frac{1}{7}, \frac{3}{7}, \frac{5}{7},$ _____, _____

c. $\frac{6}{4}, \frac{12}{4}, \frac{24}{4},$ _____, _____

SRB
51

---

5 Complete the "What's My Rule?" table and fill in the rule.

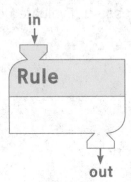

in

Rule

out

in	out
48	
40	5
1	
	3
16	2

SRB
53–54

---

6 Rewrite each whole number as a fraction with the given denominator.

a. $6 = \frac{\boxed{}}{7}$

b. $8 = \frac{\boxed{}}{8}$

c. $3 = \frac{\boxed{}}{13}$

SRB
171

232

Multiplying Mixed Numbers

For Problems 1–3:

- Use the rectangle to make an area model. Label the sides.

- Find and list the partial products. Label the partial products in the area model.

- Add the partial products to find your answer. You may need to rename fractions with a common denominator.

 $3\frac{1}{8} * 4 = ?$ Area model:

Partial products:

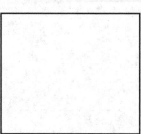

$3\frac{1}{8} * 4 =$ _____

2. $2\frac{3}{5} * \frac{1}{4} = ?$ Area model:

Partial products:

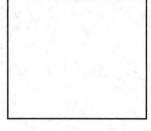

$2\frac{3}{5} * \frac{1}{4} =$ _____

3. $4\frac{1}{6} * 1\frac{2}{3} = ?$ Area model:

Partial products:

$4\frac{1}{6} * 1\frac{2}{3} =$ _____

For Problems 4 and 5, write a number model using a letter for the unknown. Then solve. Show all of your work. Remember to include units in your answer.

SRB
204-205

4 Vikram is baking bread that calls for $3\frac{3}{4}$ cups of flour. If he makes only $\frac{1}{4}$ of the recipe, how much flour will he need?

Number model: _____

Answer: _____

5 Gina is cleaning her closet and wants to know how much floor space she has. Her closet floor is a $3\frac{1}{2}$ ft by $4\frac{1}{3}$ ft rectangle. What is the area of her closet floor?

Number model: _____

Answer: _____

6 Write a number story that can be solved by multiplying $1\frac{4}{5} * 6$. Then solve.

Number story: _____

_____

_____

Answer: _____

A Hiking Trail Line Plot

Meadowbrook Nature Center has 12 hiking trails. Below is the length of each trail in miles.

$1\frac{3}{4}$ $1\frac{1}{4}$ $2\frac{1}{4}$ 2 $2\frac{1}{2}$ $1\frac{1}{2}$ $2\frac{1}{4}$ $1\frac{1}{2}$ 3   2 $2\frac{1}{2}$ $2\frac{1}{2}$

1 Complete the line plot to represent the data.

Trail Lengths

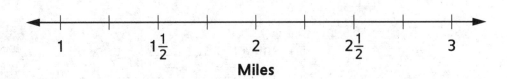

Miles

Use the line plot to answer these questions.

2 Leslie planned a family hiking trip to Meadowbrook Nature Center. On the first day her family hiked all the trails that were less than 2 miles long. How many miles did Leslie's family hike?

_____ miles

Number model: _____

3 On the second day Leslie hiked the longest trail and her little brother hiked the shortest trail. How many more miles did Leslie hike than her brother?

Leslie hiked _____ miles more than her brother.

Number model: _____

4 On the third day Leslie's father challenged himself to hike all the trails that were longer than 2 miles.

How many trails did Leslie's father hike?

_____ trails

How far did he hike?

Leslie's father hiked _____ miles

Number model: _____

Math Boxes

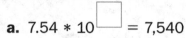

Math Boxes

---

① Write the exponent that makes each number sentence true.

a. $7.54 * 10^{\square} = 7{,}540$

b. $93.08 * 10^{\square} = 930.8$

c. $0.19 * 10^{\square} = 1{,}900$

 SRB 133

---

② Piper is buying 8 apples. Each apple costs $0.79. How much will she pay for 8 apples? Make an estimate, and then solve.

_____
(estimate)

Answer: $_____

 SRB 128, 134

---

③ Allen has 3 pears. Each pear is cut into twelfths. How many pear slices are there?

_____
(number model)

Answer: _____ slices

 SRB 209-210

---

④ Solve.

a. $10.95 + 9.028 =$ _____

b. $505.38 - 299.41 =$ _____

 SRB 130-132

---

⑤ **Writing/Reasoning** Explain the placement of the decimal point in the product in Problem 1a.

_____

_____

_____

 SRB 133

Renaming Mixed Numbers and Fractions

Math Message

SRB
171-173

Fractions greater than 1 can be written in many ways.

Example:

If is worth 1, what is   worth?

A mixed number name is $3\frac{5}{6}$. $3\frac{5}{6}$ means $3 + \frac{5}{6}$.

A fraction name is $\frac{23}{6}$. Think *sixths:*

Write the following mixed numbers as fractions. You may use your fraction circle pieces or the Fraction Number Lines Poster to help you.

1 $1\frac{2}{3} =$ _____

2 $2\frac{3}{5} =$ _____

3 $5\frac{3}{4} =$ _____

4 $6\frac{1}{5} =$ _____

5 $4\frac{7}{8} =$ _____

6 $3\frac{6}{4} =$ _____

Write the following fractions as whole numbers or mixed numbers with the greatest whole number possible. You may use your fraction circle pieces or the Fraction Number Lines Poster to help you.

7 $\frac{7}{3} =$ _____

8 $\frac{2}{1} =$ _____

9 $\frac{18}{4} =$ _____

10 $\frac{9}{3} =$ _____

11 $\frac{22}{5} =$ _____

12 $\frac{16}{8} =$ _____

Solving Mixed-Number Multiplication Problems

For Problems 1 and 2, rename any mixed numbers as fractions. Rewrite the problem using the fractions as factors. Then use a fraction multiplication algorithm to solve.

① $2\frac{3}{5} * \frac{1}{2} = ?$

② $1\frac{3}{4} * 3\frac{1}{3} = ?$

$2\frac{3}{5} * \frac{1}{2} = $ _____

$1\frac{3}{4} * 3\frac{1}{3} = $ _____

For Problems 3–5, write a number model using a letter for the unknown. Then solve using a fraction multiplication algorithm. Show your work.

③ What is the area of the sheet of notebook paper?

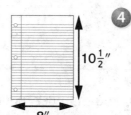

$10\frac{1}{2}''$

$8''$

_____
(number model)

④ What is the area of the flag?

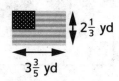

$2\frac{1}{3}$ yd

$3\frac{3}{5}$ yd

_____
(number model)

Answer: _____ in.²

Answer: _____ yd²

⑤ Last week, Aaron earned $9 doing chores. Dara earned $2\frac{1}{2}$ times as much. How much money did Dara earn?

_____
(number model)

Dara earned $_____.

⑥ Write a number story that can be solved by multiplying $1\frac{1}{2}$ and $4\frac{1}{4}$. Then solve the problem.

Number story: _____

_____

_____

Answer: _____

Math Boxes

1 Plot 4 points on the grid to form the 4 corners of a rectangle. Connect the points. List the coordinates of the 4 points below.

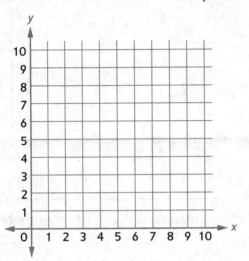

(_____, _____)　　　　　(_____, _____)

(_____, _____)　　　　　(_____, _____)

How many units from the x-axis is the base of your rectangle?

_____ units

SRB
275

---

2 Solve. Show your work.

$2\frac{3}{5} + 7\frac{1}{4} = ?$

Answer: _____

SRB
177, 191

---

3 Opal had $\frac{3}{4}$ bottle of milk. Half of Opal's milk spilled. How much milk spilled?

_____
(number model)

Answer: _____ bottle

SRB
201, 203

---

4 Oscar has 6.5 bags of topsoil. Each bag contains 0.75 cubic feet of soil. How many cubic feet of soil does Oscar have?

_____
(estimate)

Answer: _____ cubic feet

SRB
128, 134

---

5 Solve.

a. $\frac{1}{7} \div 4 =$ _____

b. $5 \div \frac{1}{10} =$ _____

SRB
207-210

239

Area Problems

Math Message

1 Find the area of the rectangle.
   Report your answer in square feet.

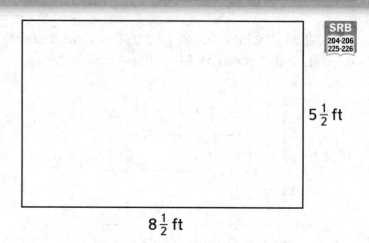

$5\frac{1}{2}$ ft

Area = _____ $8\frac{1}{2}$ ft

Solve Problems 2–5 using the method of your choice. Show your work. Write a number model to summarize each solution.

2 A typical sheet of printer paper in the United States   Work space:
   measures $8\frac{1}{2}$ inches by 11 inches.

   a. What is the area of a typical sheet of paper? _____

   Number model: _____

   b. Kim is making a design that covers $\frac{1}{4}$ square inch.
   How many times would she need to repeat her design

   to cover an entire sheet of paper? _____

   Number model: _____

3 Emerson is helping his parents order tiles for their
   bathroom floor. The area of their bathroom is 20 square
   feet. They are using tiles that are $\frac{1}{3}$ foot by $\frac{1}{3}$ foot.

   a. How many tiles are needed to cover 1 square foot?

   _____

   b. What fraction of a square foot is 1 tile? _____

   c. How many tiles are needed to cover the entire floor?

   Number model: _____

Area Problems (continued)

4 Francisco and his aunt are making a quilt. The finished quilt will be 6 ft by $7\frac{1}{2}$ ft. They are using $\frac{1}{4}$ ft by $\frac{1}{4}$ ft fabric squares.

Work space:

a. How many fabric squares will fit along the 6 ft side?

_____

b. How many fabric squares will fit along the $7\frac{1}{2}$ ft side?

_____

c. How many fabric squares will Francisco and his aunt

need to make the whole quilt? _____

Number model: _____

d. What will the area of the quilt be? _____

Number model: _____

5 Marguerite has a piece of checkered cloth like the one shown below. She measured one checker square and found that it was $\frac{1}{2}$ inch by $\frac{1}{2}$ inch. The piece of cloth has 15 rows of 20 squares each.

a. How many checker squares cover the

whole piece of cloth? _____

Number model: _____

b. What are the dimensions of the piece of

cloth? _____

c. What is the area of the cloth? _____

Number model: _____

$\Big\}\frac{1}{2}$ in.

$\frac{1}{2}$ in.

Math Boxes

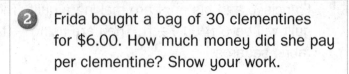
---

1 Write the exponent that makes each number sentence true.

a. $12.7 \div 10^{\square} = 0.127$

b. $2.56 \div 10^{\square} = 0.0256$

c. $107.3 \div 10^{\square} = 0.01073$

SRB
136

---

2 Frida bought a bag of 30 clementines for $6.00. How much money did she pay per clementine? Show your work.

_____
(number model)

SRB
44, 137

Answer: $ _____

---

3 Max has 4 containers of glue. He uses about $\frac{1}{6}$ container to assemble 1 robot figurine. How many robot figurines can he assemble with the glue he has?

_____
(number model)

SRB
209-210

Answer: _____ robot figurines

---

4 Solve.

a. $10.87 + 589.24 =$ _____

b. $38.2 - 6.017 =$ _____

SRB
130-132

---

5 **Writing/Reasoning** Draw a picture that shows how you solved Problem 3.

SRB
209

Using Common Denominators to Divide

Math Message

Use your fraction circle pieces to solve $4 \div \frac{1}{3}$. Draw lines on the circles below to show how you solved it.

$4 \div \frac{1}{3} =$ _____

Using Common Denominators to Divide

One way to divide fractions is to use common denominators. This method can be used to divide whole numbers by fractions and fractions by whole numbers.

Step 1 Rename the dividend and divisor as fractions with a common denominator.

Step 2 Divide the numerators.

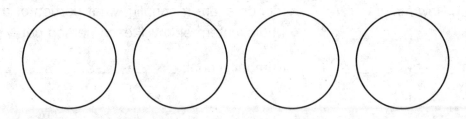

$$\textbf{Examples:} \quad 3 \div \frac{1}{2} = \frac{6}{2} \div \frac{1}{2} \qquad\qquad \frac{1}{3} \div 4 = \frac{1}{3} \div \frac{12}{3}$$

$$= 6 \div 1 \qquad\qquad\qquad\quad = 1 \div 12$$

$$= 6 \qquad\qquad\qquad\qquad\quad = \frac{1}{12}$$

Solve. Show your work. Use multiplication to check your answers.

1 $2 \div \frac{1}{5} = ?$

2 $\frac{1}{4} \div 4 = ?$

Answer: _____

Answer: _____

Check: _____

Check: _____

3 $\frac{1}{2} \div 5 = ?$

4 $4 \div \frac{1}{4} = ?$

Answer: _____

Answer: _____

Check: _____

Check: _____

Fraction Division Problems

For Problems 1–4, write a number model using a letter for the unknown. Solve using the strategy of your choice. Show your work. Use multiplication to check your answers.

SRB
207-210

1 Chase is packing flour in $\frac{1}{2}$-pound bags. He has 8 pounds of flour. How many bags can he pack?

Number model: _____

Answer: _____ bags

Check: _____

2 Regina has $\frac{1}{4}$ watermelon. If she and two friends share it equally, what portion of the whole watermelon will each person get?

Number model: _____

Answer: _____ watermelon

Check: _____

3 Four students are running a relay race that is $\frac{1}{2}$ mile long. Each student runs the same distance. How far does each student run?

Number model: _____

Answer: _____ mile

Check: _____

4 A cook baked 5 large pizzas. If he cuts each pizza into eighths, how many slices will he have?

Number model: _____

Answer: _____ slices

Check: _____

For Problems 5 and 6, write a number story that can be modeled by the expression. Then solve.

5 $3 \div \frac{1}{4}$

Number story: _____

_____

_____

Solution: _____

6 $\frac{1}{3} \div 4$

Number story: _____

_____

_____

Solution: _____

244

1 Plot 3 points on the grid to form the 3 vertices of a triangle. Connect the points. List the coordinates of the 3 points below.

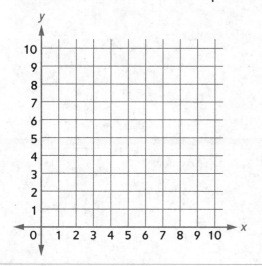

(____, ____)

(____, ____)

(____, ____)

Circle one of your ordered pairs. How many units away from the *y*-axis is the point that you circled? _____ units

SRB
275

2 Solve. Show your work.

$5\frac{3}{4} + 1\frac{2}{3} = ?$

Answer: _____

SRB
177, 191

3 Danielle spent $\frac{2}{5}$ of her paycheck on bills. Her water bill was $\frac{1}{6}$ of what she spent on bills. What part of her paycheck went to her water bill?

_____
(number model)

Answer: _____

SRB
201, 203

4 Bill has a summer job that pays $9.60 per hour. What is his total pay for 71.5 hours of work?

_____
(estimate)

Answer: $_____

SRB
128, 134

5 Solve.

a. $1 \div \frac{1}{11} =$ _____

b. $\frac{1}{7} \div 7 =$ _____

SRB
207-210

A Triangle Hierarchy

On the left, write the categories and subcategories from the triangle hierarchy you create in class. Use the hierarchy to classify your triangle cards. When you are finished, glue or tape the cards in place.

Triangle Hierarchy   **Triangle Cards**

A Triangle Hierarchy (continued)

Lesson 7-5

DATE   TIME

Use the triangle hierarchy on journal page 246 to answer the questions.

1 Name a category shown on the hierarchy. Then name a subcategory of that category.

   a. Category: _____

   b. Subcategory: _____

2 **a.** Draw an isosceles triangle.   **b.** Draw a triangle that is *not* an isosceles triangle.

3 **a.** Draw an equilateral triangle.   **b.** Is your equilateral triangle also an isosceles triangle? _____

                                       Explain. _____

                                       _____

                                       _____

                                       _____

4 In each statement below, replace the underlined category with a different category from the triangle hierarchy so that the new statement is still true.

   a. All <u>triangles</u> have three sides and three angles.

      All _____ have three sides and three angles.

   b. All <u>isosceles triangles</u> have at least two sides the same length.

      All _____ have at least two sides the same length.

   c. All <u>isosceles triangles</u> have a line of symmetry.

      All _____ have a line of symmetry.

5 Look at your answers to Problem 4. Describe any patterns that you see.
Hint: Think about categories and subcategories.

_____

_____

247

Math Boxes

---

1 Find the area of the rectangle.

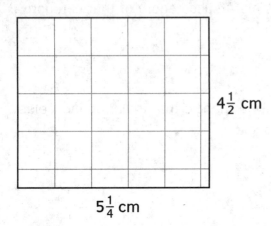

$4\frac{1}{2}$ cm

$5\frac{1}{4}$ cm

Area: _____

_____
(number model)

SRB
204-205
225

---

2 Solve.

a. $\frac{1}{7} * \frac{8}{9} =$ _____

b. $\frac{6}{11} * \frac{3}{4} =$ _____

SRB
203

---

3 Use an estimate to place the decimal point. The correct digits are given.

_____
(estimate)

$2.3 * 158.6 = 3 \quad 6 \quad 4 \quad 7 \quad 8$

SRB
128, 134

---

4 Fill in each blank with <, >, or =.

a. $12 * \frac{1}{3}$ _____ 12

b. $\frac{2}{3} * \frac{8}{8}$ _____ $\frac{2}{3}$

c. $1\frac{3}{8} * \frac{6}{5}$ _____ $1\frac{3}{8}$

SRB
197-198

---

5 **Writing/Reasoning** Explain how you solved Problem 2b.

_____

_____

_____

_____

SRB
203

Solving Bulletin Board Area Problems

Solve the problems below. Write a number model to summarize each solution.

SRB
204-206
225-226

1. The bulletin board in Shain's classroom is $3\frac{1}{4}$ feet by $4\frac{1}{2}$ feet.

   a. What is the area of the bulletin board?

   Area: _____   Number model: _____

   b. Shain is going to use a piece of blue fabric to cover the bulletin board. The fabric is 7 feet by $3\frac{1}{2}$ feet. What is the area of the fabric?

   Area: _____   Number model: _____

   c. How many square feet of fabric will Shain have left after he covers the bulletin board?

   Answer: _____   Number model: _____

2. The bulletin board in Marlon's classroom is $4\frac{1}{3}$ feet by 8 feet.

   a. What is the area of the bulletin board?

   Area: _____   Number model: _____

   b. Whenever Marlon and her classmates finished reading a book, they wrote the title on a $\frac{1}{3}$ ft by $\frac{1}{3}$ ft piece of colored paper. Each piece of paper was stapled to the bulletin board, side by side, without overlapping. By the end of the school year, the entire bulletin board was covered with book titles. How many pieces of paper were on the bulletin board?

   Answer: _____   Number model: _____

A Quadrilateral Hierarchy

Math Message

Work with a partner. Carefully read each of the definitions below. Then find one quadrilateral card that shows an example of each type of quadrilateral. Write the letter of your example next to the definition.

_____ A **trapezoid** is a quadrilateral that has at least one pair of parallel sides.

_____ A **kite** is a quadrilateral with two separate pairs of adjacent equal-length sides.

_____ A **parallelogram** is a trapezoid with two pairs of parallel sides.

_____ A **rhombus** is a parallelogram with all four sides equal in length.

_____ A **rectangle** is a parallelogram with four right angles.

_____ A **square** is a rectangle with all four sides equal in length.

Use the hierarchy on journal page 251 to classify your quadrilateral cards. Then use the hierarchy to answer the questions.

1 From the hierarchy you can tell that all trapezoids are quadrilaterals, but not all quadrilaterals are trapezoids.

Use the hierarchy to write two other statements like this.

a. All _____ are _____, but not all _____ are _____.

b. All _____ are _____, but not all _____ are _____.

2 The relationships shown in the hierarchy also help you think about properties.

For example, all parallelograms have two pairs of parallel sides, and all rectangles are parallelograms, so all rectangles have two pairs of parallel sides.

a. Use the hierarchy to help you complete this statement:

All kites have two pairs of equal-length sides that are next to each other, and all

rhombuses are kites, so all rhombuses _____

_____

b. Use the hierarchy to help you write one more statement like the one in Part a.

_____

_____

_____

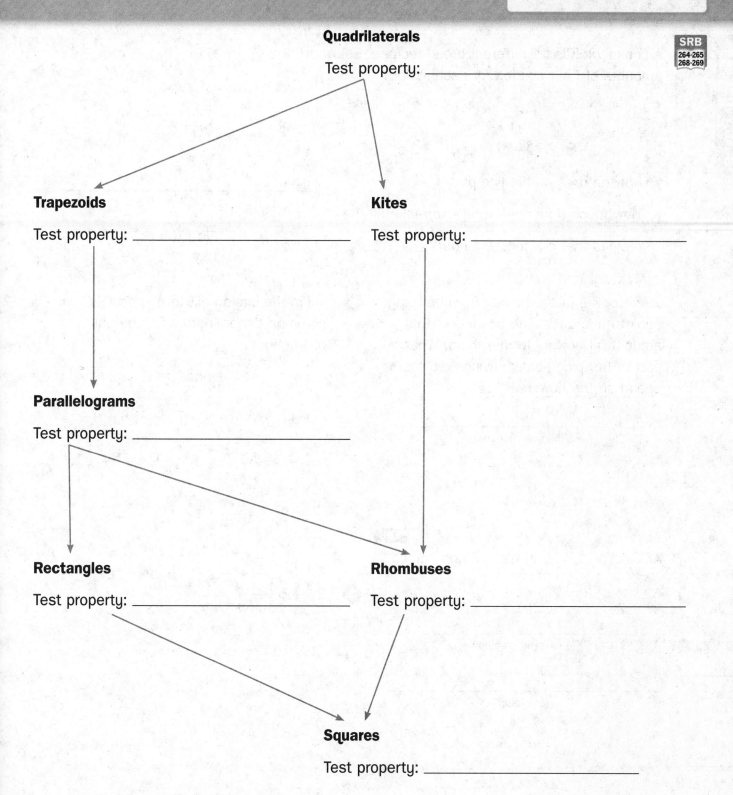

Quadrilaterals

Test property: _____

Trapezoids

Test property: _____

Kites

Test property: _____

Parallelograms

Test property: _____

Rectangles

Test property: _____

Rhombuses

Test property: _____

Squares

Test property: _____

Math Boxes

1 A cookbook has 11 different bread recipes. Below are the amounts of flour needed for each recipe.

$1\frac{1}{2}$ c $2\frac{1}{4}$ c $2\frac{1}{2}$ c   2 c   2 c $1\frac{3}{4}$ c

$2\frac{1}{4}$ c $1\frac{3}{4}$ c $2\frac{1}{4}$ c   2 c   2 c

a. Put the data on the line plot.

b. How much flour does it take to make all the recipes that call for $2\frac{1}{4}$ c flour? _____

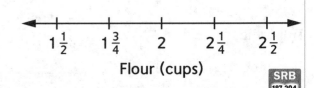

Flour (cups)

SRB
187, 204,
244, 247

---

2 Josie took $\frac{1}{3}$ of the money from her piggy bank to buy gifts. She spent $\frac{2}{5}$ of that amount on flowers for her mom. What part of her piggy-bank savings did Josie spend on the flowers?

_____
(number model)

Answer: _____ of her savings

SRB
201, 203

3 Fill in the circle next to <u>all</u> possible common denominators for the pair of fractions.

$\frac{1}{3}$ and $\frac{5}{9}$

◯ **A.** 9   ◯ **B.** 24

◯ **C.** 36   ◯ **D.** 18

◯ **E.** 96

SRB
177

---

4 62.1 ÷ 2.3 = ?

_____
(equivalent problem)

_____
(estimate)

62.1 ÷ 2.3 = _____

SRB
128,
140-142

5 $2\frac{3}{5} * 6\frac{1}{4} = ?$

Answer: _____

SRB
204-206

Solve Problems 1–3 using partial products. Show your work.

1 $5 * 3\frac{1}{6} = ?$

2 $4\frac{3}{4} * \frac{1}{2} = ?$

$5 * 3\frac{1}{6} =$ _____

$4\frac{3}{4} * \frac{1}{2} =$ _____

3 The rug in Mr. Flint's classroom is $7\frac{1}{4}$ ft by $5\frac{1}{2}$ ft. What's the area of the rug?

Number model: _____

Answer: _____

For Problems 4–6, rename each factor as a fraction and use a fraction multiplication algorithm.

4 $2\frac{5}{8} * \frac{1}{3} = ?$

5 $3\frac{4}{5} * 2 = ?$

$2\frac{5}{8} * \frac{1}{3} =$ _____

$3\frac{4}{5} * 2 =$ _____

6 A jumbo bag of popcorn kernels contains $2\frac{1}{2}$ times as much as a small bag. If a small bag has $10\frac{1}{2}$ ounces of popcorn kernels, how many ounces of kernels are in the jumbo bag?

Number model: _____

Answer: _____

7 Compare the two methods for multiplying mixed numbers. Which do you prefer? Why?

_____

_____

_____

Math Boxes

Math Boxes

---

① Find the area of the rectangle.

$\frac{1}{2}$ unit

$\frac{1}{2}$ unit

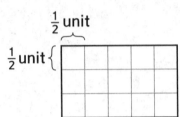

Area: _____

_____
(number model)

---

② Solve.

a. $\frac{7}{9} * \frac{3}{5} =$ _____

b. $\frac{5}{11} * \frac{6}{9} =$ _____

---

③ Use an estimate to place the decimal point. The correct digits are given.

_____
(estimate)

$6.39 \div 21.3 = 0 \quad 0 \quad 3 \quad 0$

---

④ Which expressions have a value equal to 6? Check all that apply.

☐ $6 * \frac{2}{2}$

☐ $6 * \frac{8}{7}$

☐ $\frac{3}{2} * \frac{6}{1}$

☐ $6 * \frac{9}{10}$

☐ $6 * 1$

---

⑤ **Writing/Reasoning** Explain how you solved Problem 4 without multiplying.

_____

_____

_____

_____

Geometric Attributes

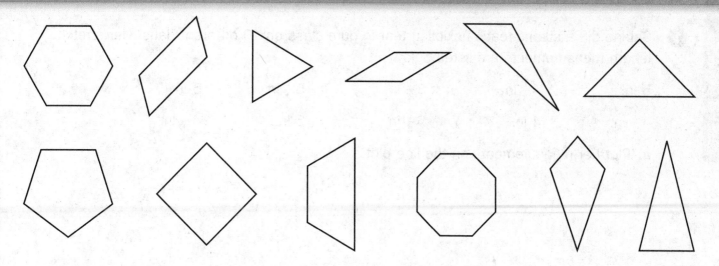

Name at least three attributes that all of the shapes have in common.

Math Boxes

1 Sabine did a sit-and-reach flexibility test in gym class on 10 different days. Her stretch length measurements are listed below.

5 in.	$4\frac{1}{4}$ in.	$3\frac{1}{4}$ in.	$5\frac{1}{2}$ in.	5 in.
$4\frac{1}{2}$ in.	4 in.	$5\frac{1}{4}$ in.	5 in.	$4\frac{3}{4}$ in.

a. Plot her measurements on the line plot.

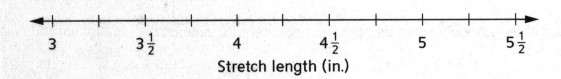

3       $3\frac{1}{2}$ 4 $4\frac{1}{2}$ 5 $5\frac{1}{2}$

Stretch length (in.)

b. What is the difference between Sabine's highest and lowest measurements?

_____ in.

SRB
192,
244, 247

---

2 On Monday morning, Elijah opened a box of cereal and ate $\frac{1}{8}$ of its contents. On Tuesday, he ate $\frac{1}{6}$ of what was left. How much did he eat on Tuesday?

_____
(number model)

Answer: _____ box of cereal

SRB
201, 203

3 Solve.

a. $\frac{7}{8} + \frac{3}{4} = $ _____

b. $\frac{7}{9} + \frac{1}{2} = $ _____

SRB
177,
189-190

---

4 $72.5 \div 14.5 = ?$

_____
(equivalent problem)

_____
(estimate)

$72.5 \div 14.5 = $ _____

SRB
128,
140-142

5 $6\frac{3}{8} * 4\frac{1}{3} = ?$

Answer: _____

SRB
204-206

Math Boxes

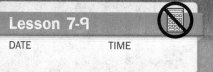

Math Boxes

1 Justin said, "If I multiply any number greater than 0 by $\frac{6}{5}$, the product will be smaller than the number I started with." Is Justin's conjecture true or false?

_____

SRB
197-198

2 Solve. Check your answer with a multiplication number sentence.

$12 \div \frac{1}{7} = $ _____

Check: _____

SRB
209-210

3 You can use this formula to convert temperature from degrees Celsius (°C) to degrees Fahrenheit (°F):

$(°C * 1.8) + 32 = °F$

What is 31.5°C in °F?

Answer: _____ °F

SRB
42,
134-135

4 Solve the following riddle:
I am a quadrilateral with 4 equal sides and no right angles. What could I be? Fill in all possible answers.

 square

( ) parallelogram

( ) rhombus

( ) kite

SRB
268-269

5 **Writing/Reasoning** Explain how you decided whether Justin's conjecture was true or false for Problem 1.

_____

_____

_____

SRB
197-198

Finding Personal Measures

Work with a partner. Use a tape measure to find the length of your partner's fathom, cubit, great span, and joint in standard units. Record your own data on your journal page. Your partner's data should be recorded in his or her journal. Be sure to measure to the level of precision listed.

SRB
244-245
248

1 Fathom
(to the nearest inch)

2 Cubit
(to the nearest $\frac{1}{2}$ inch)

3 Great span
(to the nearest $\frac{1}{4}$ inch)

4 Joint
(to the nearest $\frac{1}{8}$ inch)

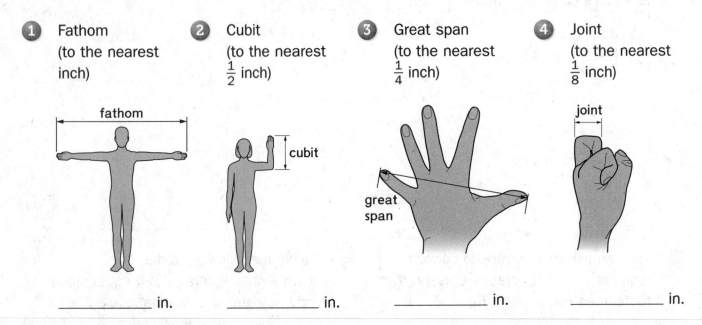

_____ in.   _____ in.   _____ in.   _____ in.

5 Use your class measurements to create line plots for the cubit lengths, great-span lengths, and joint lengths.

Cubit Length (in.)

Great-Span Length (in.)

Joint Length (in.)

6 Use the evening-out strategy to find a typical measurement for your class.
Remember: To even out a data set, add all of the values and then divide the total
by the number of data points.

a. Great-span data _____   **b.** Joint data _____

7 How much longer is the typical great span than the typical joint?

Number model: _____   Answer: _____

8 If everyone in your class had the typical great-span measurement and you *all* lined up one
of your hands thumb to pinkie, what would the total distance be?

Number model: _____   Answer: _____

Visualizing Patterns and Relationships

 a. Write the numbers in the table as ordered pairs where the *in* numbers are the *x*-coordinates and the *out* numbers are the *y*-coordinates.

in (x)	out (y)
Rule: + 1	Rule: + 2
1	2
2	4
3	6
4	8
5	10

in/out relationship rule: * 2

Ordered pairs:

(1, 2)

_____

_____

_____

_____

b. Graph the ordered pairs on the coordinate grid. Then draw a line to connect the points.

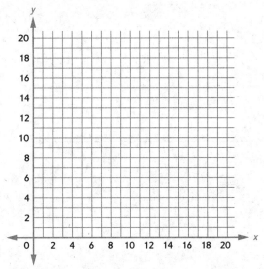

 a. A rule is given at the top of each column in the table below. Use the rules to fill in the columns.

in (x)	out (y)
Rule: + 6	Rule: + 2
0	0

Ordered pairs:

_____

_____

_____

_____

_____

d. Write the numbers from the table as ordered pairs. Then graph them. Draw a line to connect the points.

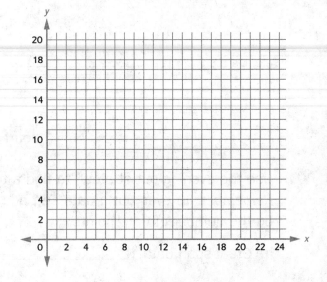

b. What rule relates each *in* number to its corresponding *out* number? _____

c. Think about the rules you used to fill in the *in* and *out* columns. Why does the rule you found in Part b make sense?

_____

_____

_____

Visualizing Patterns and Relationships (continued)

SRB
51-56
275

3 **a.** Use the rules to fill in the columns.

in (x) Rule: − 1	out (y) Rule: − 3
5	15

b. What rule relates each *in* number to its corresponding *out* number?

_____

c. Write the numbers from the table as ordered pairs. Then graph the ordered pairs. Draw a line to connect the points.

Ordered pairs: _____

_____

_____

_____

_____

4 **a.** Use the rules to fill in the columns.

in (x) Rule: − 4	out (y) Rule: − 2
20	10

b. What rule relates each *in* number to its corresponding *out* number?

_____

c. Write the numbers from the table as ordered pairs. Then graph the ordered pairs. Draw a line to connect the points.

Ordered pairs: _____

_____

_____

_____

_____

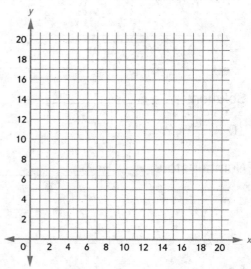

Multiplying and Dividing Fractions

For Problems 1–4, write a number model with a letter for the unknown. Then solve. You may draw pictures to help you.

1 Tony made a painting that is $\frac{3}{4}$ ft wide. He wants to make a new painting that is $\frac{1}{2}$ as wide. What will be the width of the new painting?

Number model: _____

Answer: _____ ft

2 Rachel is decorating a card with stickers. Each sticker is $\frac{1}{4}$ in. across. How many stickers can Rachel fit along the 5 in. edge of the card?

Number model: _____

Answer: _____ stickers

3 A Chicago city block is about $\frac{1}{8}$ mile long. If the length of a block is divided into 10 equal lots, what is the length of one lot in miles?

Number model: _____

Answer: _____ mile

4 Greentown's city council is designing a new park. The park will be $\frac{1}{4}$ mile long and $\frac{1}{8}$ mile wide. How many square miles will the park cover?

Number model: _____

Answer: _____ square mile

For Problems 5 and 6, write a number story to match the number sentence. Then solve.

5 $\frac{3}{4} * \frac{5}{8} = ?$

Number story: _____

_____

_____

Solution: _____

6 $10 \div \frac{1}{3} = ?$

Number story: _____

_____

_____

Solution: _____

Math Boxes: Preview for Unit 8

1 Adam has a couch that is 7 feet long and two end tables that are each 18 inches wide. If he puts one end table at each end of the couch, how many **feet** of space will the furniture take up?

Answer: _____

SRB
215-217
328

2 Carlene is working on a rectangular puzzle that is $\frac{11}{12}$ foot by $1\frac{2}{3}$ feet. What is the area of the completed puzzle?

_____
(number model)

Answer: _____ ft²

SRB
204-206

3 Write whether you would find length, area, or volume for each situation.

a. The amount of water in a pool

_____

b. The distance from your school to your

home _____

c. The amount of paper needed to cover

a bulletin board _____

SRB
218, 221,
230

4 On the lines below, write the value of the digits in 6,582,390,417 in word form.

6: _____

5: _____

8: _____

2: _____

3: _____

SRB
66-67

5 Complete the in/out table. Then complete the rule that relates the *in* numbers to the corresponding *out* numbers.

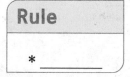

in	out
0	
1	$\frac{1}{4}$
2	
	$\frac{1}{8}$
8	2

Rule

* _____

SRB
53-54

263

Displaying Data in a Table and on a Graph

Follow your teacher's directions to complete Problem 1.

SRB
52-56

 a.

Time (minutes) (x)	Distance (miles) (y)
0	0
1	8
2	16
3	24
4	32
5	

b. Rule: _____

_____

c. Ordered pairs:

_____

_____

_____

_____

_____

_____

2 Plot the ordered pairs on the coordinate grid on journal page 265. Connect the points with a straightedge.

Use your graph to answer the questions below.

3 How far did the plane travel in $2\frac{1}{2}$ minutes? _____
(unit)

4 About how many miles did the plane travel in 5 minutes 24 seconds ($5\frac{2}{5}$ minutes)?

_____
(unit)

5 About how long did it take the plane to travel 60 miles? _____
(unit)

6 How long did it take the plane to travel 64 miles? _____
(unit)

Displaying Data in a Table and on a Graph (continued)

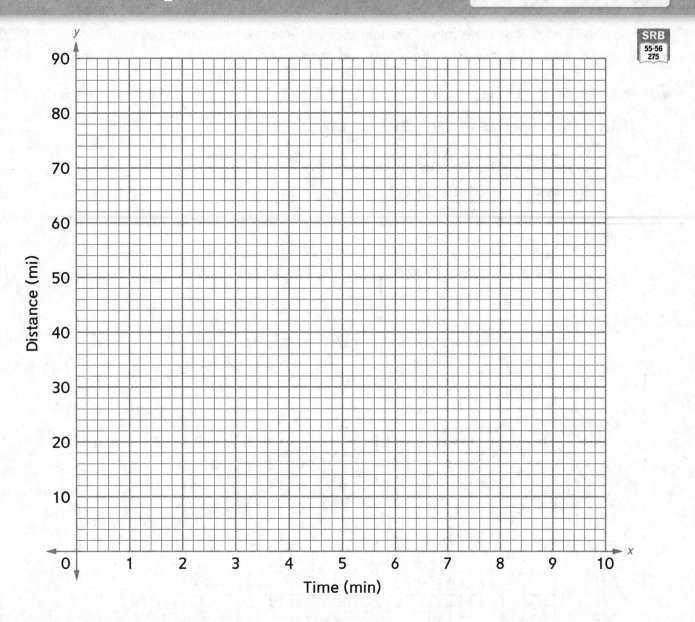

7 If you want to know how far the plane traveled in 15 minutes, would you use the table, the rule, or the graph? Explain.

_____

_____

_____

Creating Graphs from Tables

For Problems 1 and 2, use the rule to complete each table. Write ordered pairs to represent the data. Then create a line graph and answer the questions.

1. Andy earns $8 per hour.

   Rule: Number of hours worked * $8 = Earnings

   a.

Hours Worked (x)	Earnings (dollars) (y)
1	
2	
3	
	40
7	

   b. Ordered pairs:

   _____

   _____

   _____

   _____

   _____

   c.

   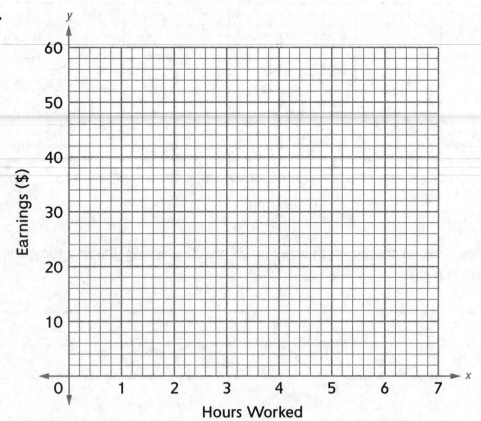

   d. Plot a point to show Andy's earnings for $5\frac{1}{2}$ hours. How much would he earn? _____

266

Creating Graphs from Tables (continued)

2 Frank types 45 words per minute.

SRB
55-56
275

Rule: Words typed = 45 * number of minutes

a.

Time (minutes) (x)	Words Typed (y)
1	
2	
3	
	225
6	

b. Ordered pairs:

_____

_____

_____

_____

_____

c.

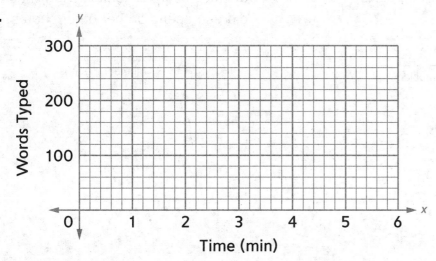

d. Plot a point to show how many words Frank could type in 4 minutes. How many words

could he type? _____
                     (unit)

e. About how long would it take Frank to type 200 words? _____
                                                                (unit)

Math Boxes

1 Use a common denominator to divide. Show your work.

   a. $7 \div \frac{1}{4} =$ _____

   b. $5 \div \frac{1}{6} =$ _____

SRB
210

2 Solve. Show your work.

$4\frac{1}{2} * 7\frac{1}{5} = ?$

Answer: _____

SRB
204-206

3 Write an equivalent problem with a whole-number divisor. Then solve.

$15.6 \div 0.3 = ?$

_____
(equivalent problem)

Answer: _____

SRB
140-141

4 Jorie spent $1\frac{3}{4}$ hours doing chores. She spent $\frac{2}{3}$ hour vacuuming. How much time did she spend on her other chores?

_____
(number model)

Answer: _____ hours

SRB
192-193

5 **Writing/Reasoning** Explain how you solved Problem 3.

_____

_____

_____

SRB
140-141

Math Boxes

① Write a fraction to make each number sentence true.

a. $\frac{6}{11} * \dfrac{\boxed{}}{\boxed{}} = \frac{6}{11}$

b. $\frac{6}{11} * \dfrac{\boxed{}}{\boxed{}} > \frac{6}{11}$

c. $\frac{6}{11} * \dfrac{\boxed{}}{\boxed{}} < \frac{6}{11}$

SRB 197-198

② Write a *division* number sentence that matches the picture below. Your dividend should be a whole number and your divisor should be a fraction.

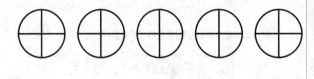

_____ ÷ _____ = _____

SRB 209-210

③ Noah bought 8.5 gallons of gas. Gas is $3.90 per gallon. How much did Noah pay for gas?

_____
(estimate)

Answer: $_____

SRB 128, 134-135

④ What are all the possible names for a polygon with 4 right angles and 4 equal sides?

Circle the best answer.

A. Rectangle and square

B. Square

C. Quadrilateral, rhombus, kite, rectangle, square, trapezoid, parallelogram

D. Parallelogram and square

SRB 268-269

⑤ **Writing/Reasoning** Explain how you found your answer to Problem 1b.

_____

_____

_____

SRB 197-198

269

Math Message

SRB
55-56

1 Karla earns $20 per hour.

   a. Write a rule that describes how much Karla earns.

   Rule: _____

   b. Use the rule to complete the table. Write ordered pairs to represent the data in the table.

Hours Worked (x)	Earnings (dollars) (y)
1	
2	
3	

Ordered pairs:

_____

_____

_____

---

2 Eli is 10 years old and can run about 5 yards in 1 second. His sister Lupita is 7 years old and can run about 4 yards in 1 second.

Eli and Lupita have a 60-yard race. Because Lupita is younger, Eli gives her a 10-yard head start.

Complete the tables showing the distances Eli and Lupita are from the starting line after 1 second, 2 seconds, 3 seconds, and so on. Use the tables to complete Problem 2.

   a. Who wins the race? _____

   b. How long does it take the winner to finish? _____

   c. Who was in the lead during most of the race? _____

Eli		Lupita	
Time (seconds) (x)	Distance (yards) (y)	Time (seconds) (x)	Distance (yards) (y)
0 (Start)	0	0 (Start)	10
1		1	
2		2	18
3	15	3	
4		4	
5		5	
6		6	
7		7	38
8		8	
9		9	
10		10	
11		11	
12		12	

3   **a.** Eli's rule: _____

   b. Lupita's rule: _____

Rules, Tables, and Graphs (continued)

4 Write 4 ordered pairs from Eli's table and 4 ordered pairs from Lupita's table.

SRB
55-56
275

a. Eli: _____

b. Lupita: _____

5 Use the grid below to graph the results of the race between Eli and Lupita.

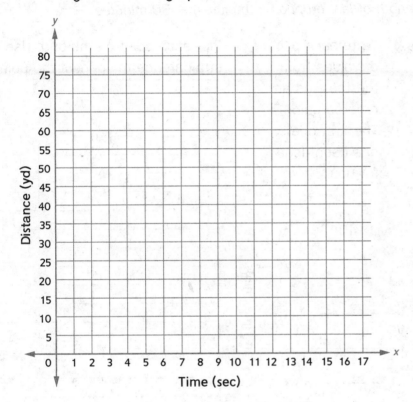

6 How many yards apart are Eli and Lupita at 7 seconds? _____ yards

7 Suppose Eli and Lupita race for 75 yards instead of 60 yards.

a. Who would you expect to win? _____

b. How long would it take the winner to finish? _____ seconds

8 Explain how you figured out the answers to Problems 7a and 7b.

_____

_____

_____

_____

_____

271

Predicting Old Faithful's Eruptions

Old Faithful, one of 200 geysers found in Yellowstone National Park, is one of nature's most impressive sights. Old Faithful is not the largest or tallest geyser in Yellowstone, but it *is* the most dependable. It erupts at predictable intervals. If you time the length of one eruption, you can predict how long you will have to wait until the next eruption. The formula that describes Old Faithful's eruption pattern is:

Wait time = (10 * length of last eruption in minutes) + 30 minutes

1 Use the formula to complete the table.

Length of Previous Eruption (minutes) (x)	Wait Time to Next Eruption (minutes) (y)
1	40
2	
3	
4	
5	
6	
$2\frac{1}{2}$	
	35

2 Write each pair of values from the table as an ordered pair. The first one has been done for you.

(1, 40)

_____

_____

_____

_____

_____

_____

_____

3 Plot each point on the grid. Use a straightedge to connect the points.

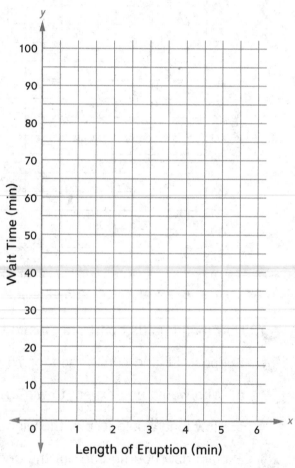

4 a. It is 8:30 A.M. and Old Faithful has just finished a 4-minute eruption. At about what time will the next eruption occur? _____

b. Explain how you found your answer.

_____

_____

272

Math Boxes

 Math Boxes

---

1 Use a common denominator to divide.

a. $\frac{1}{3} \div 5 = $ _____

b. $\frac{1}{7} \div 2 = $ _____

 SRB
210

2 Solve. Show your work.

$4\frac{1}{6} * 1\frac{7}{8} = ?$

SRB
204-206

Answer: _____

---

3 Write an equivalent problem with a whole number divisor. Then solve.

$194.5 \div 0.5 = ?$

_____
(equivalent problem)

SRB
140-141

Answer: _____

4 A marathon is approximately $26\frac{1}{5}$ miles. If Marlo has already run $17\frac{1}{3}$ miles, how much farther does she have to go?

_____
(number model)

 SRB
192-193

Answer: _____ miles

---

5 **Writing/Reasoning** Complete the area model to show that your answer to Problem 2 is correct.

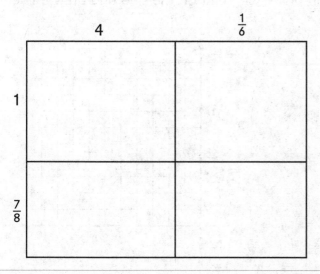

SRB
204-205

Math Boxes:
Preview for Unit 8

Math Boxes

---

① **a.** 1 hour = _____ seconds

   b. 1 day = _____ seconds

SRB
215-217
328

---

② The dimensions of a standard basketball court are $31\frac{1}{3}$ yards by $16\frac{2}{3}$ yards. What is the area of a basketball court?

_____
(number model)

Answer: _____ yd²

SRB
204-206

---

③ Write about a situation in which you might want to know the volume of a rectangular prism.

_____

_____

_____

SRB
230

---

④ Write a number with 7 ten-millions, 2 billions, 9 hundred-thousands, 3 ones, and 5 thousands. Write a 4 in all the other places.

____ , ____ ____ ____ , ____ ____ ____ , ____ ____ ____

Write this number in words: _____

_____

_____

SRB
66-67

---

⑤ Use the table to write ordered pairs. Then plot each point on the grid and draw a line to connect them.

in	out
0	0
1	$\frac{1}{4}$
2	$\frac{2}{4}$
3	$\frac{3}{4}$
4	1

_____

_____

_____

_____

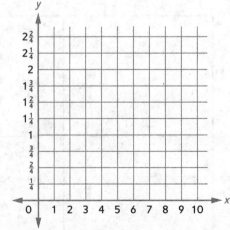

SRB
275

Finding Areas of Playing Surfaces

A town is planning to build an athletic center on some unused land. The planners are trying to figure out the areas of playing surfaces for a variety of sports. Help the planners by finding the area of each playing surface in square feet. Show your work on the grid below. Use another sheet of paper if you need more space.

Sport	Dimensions of Playing Surface	Dimensions of Playing Surface in Feet	Area of Playing Surface in Square Feet
Beach volleyball	52 ft 6 in. × 26 ft 3 in.	_____ ft × _____ ft	
Karate	26 ft × 26 ft	_____ ft × _____ ft	
Judo	52 ft 6 in. × 52 ft 6 in.	_____ ft × _____ ft	
Football (U.S., without end zones)	100 yd × $53\frac{1}{3}$ yd	_____ ft × _____ ft	
Soccer	120 yd × 80 yd	_____ ft × _____ ft	
Swimming	55 yd × 23 yd	_____ ft × _____ ft	
Ice hockey	30 m × 61 m (1 m ≈ $\frac{12}{11}$ yd)	_____ ft × _____ ft	About
Wrestling	39 ft 3 in. × 39 ft 3 in.	_____ ft × _____ ft	

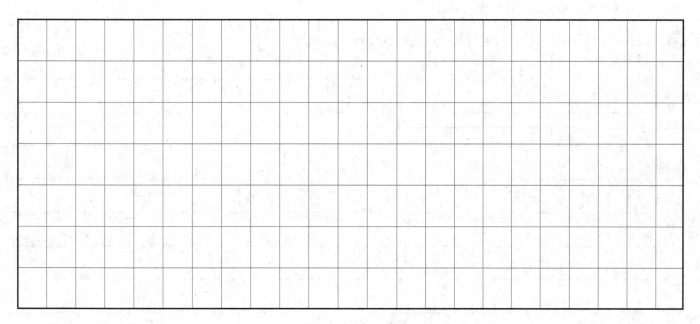

Planning an Athletic Center

SRB
215-217

1   An acre is 4,840 square yards. How many square feet is that?

Answer: _____

The town has 4 acres of land to use for an athletic center.
The land is a rectangle with a length of 160 yards and a width of 121 yards.

The town wants the athletic center to have a variety of sports playing surfaces. You have been asked to help decide which playing surfaces should be included and how the surfaces should be arranged.

Use the following guidelines to help you plan the athletic center.

• Choose from the sports playing surfaces listed on journal page 275. Use the dimensions you calculated on that page to help you plan.

• As you arrange the playing surfaces, remember that there should be some space between the playing surfaces for walkways and for spectators.

• You can use extra space for features like a warm-up area or a snack stand.

• Use blank pieces of paper to try out different plans. When you have a final plan, draw it on journal page 277. Be sure to label each playing surface with the name of the sport, the dimensions, and the area.

2   Explain how you and your group created your plan.

_____

_____

_____

_____

_____

_____

_____

Draw and label your plan for the athletic center on this page.

Math Boxes

1 Choose the answer that makes the following sentence true.

When you multiply a number by a fraction less than 1, the product is _____

○ less than the original number.

○ equal to the original number.

○ greater than the original number.

SRB
198

2 Joakim is buying movie tickets for himself and 3 friends for $9.50 each. How much will the tickets cost all together?

_____
(number model)

Answer: $ _____

SRB
44,
134-135

3 What is the area of a rectangular park that is $\frac{5}{8}$ mile long and $\frac{2}{5}$ mile wide?

_____
(number model)

Answer: _____

SRB
202-203
225

4 Delaney is cutting pita rounds into eighths. If she has 6 pita rounds, how many pieces of pita will she have?

_____
(division number model)

Answer: _____

SRB
209-210

5 **Writing/Reasoning** Explain how you chose your answer for Problem 1.

_____

_____

_____

_____

_____

SRB
198

278

Finding Areas of Nonrectangular Figures

Math Message

Find the area of the rectangle in Problem 1a. Then talk with a partner about how you could find the area of the triangle in Problem 1b.

1 **a.**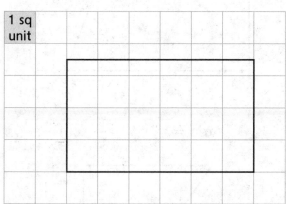

b.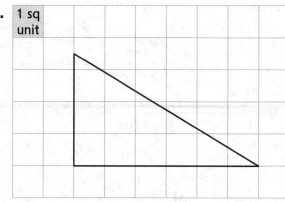

Area: _____ units²

Area: _____ units²

---

2 Explain how you used the rectangle in Problem 1a to help you find the area of the triangle in Problem 1b.

_____

_____

_____

_____

_____

3 Explain the **rectangle method** for finding area.

_____

_____

_____

_____

_____

Using the Rectangle Method to Find Area

Use the rectangle method to find the area of each figure in square centimeters.

SRB
228-229

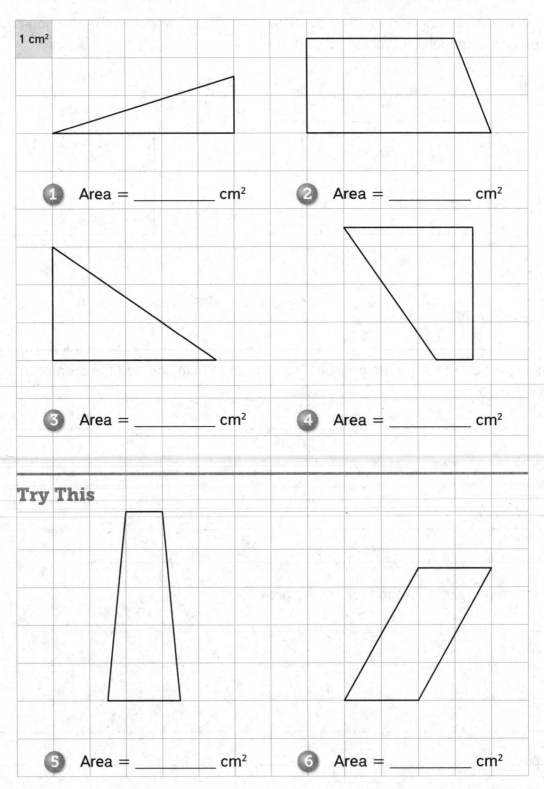

1 cm²

1 Area = _____ cm²

2 Area = _____ cm²

3 Area = _____ cm²

4 Area = _____ cm²

Try This

5 Area = _____ cm²

6 Area = _____ cm²

Organic apples are on sale for $1.50 per pound. Anita started making the table in Problem 1 to show how much different amounts of apples will cost. She thinks of $1.50 as 1.5 dollars. **SRB** 55-56 275

She says, "Each time I add 1 pound of apples, I add another $1.50 to the total cost."
She writes the rule "+ 1" in the Pounds of Apples column and the rule "+ 1.5" in the Total Cost in Dollars column.

1 Use the rules to complete Anita's table.

Pounds of Apples (x) Rule: + 1	Total Cost in Dollars (y) Rule: + 1.5
1	1.5

2 **a.** Look at the table in Problem 1. What rule relates the Pounds of Apples column to the Total Cost in Dollars column?

Rule: _____

b. Why does the rule you found in Part a make sense?

_____

_____

_____

_____

3 Write the numbers from the table in Problem 1 as ordered pairs. Remember to use parentheses and a comma in each ordered pair. Then graph the ordered pairs on the coordinate grid. Draw a line to connect the points.

Ordered pairs: _____

_____   _____

_____   _____

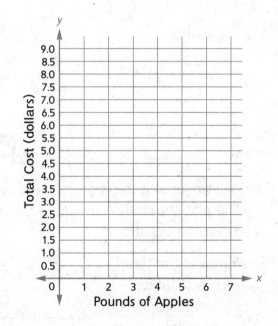

4 Use the table. How much will 3 pounds of apples cost? _____

5 Use the rule from Problem 2a. How much will 10 pounds of apples cost? _____

6 Use the graph. How much will 6 pounds of apples cost? _____

Math Boxes

1 Use an estimate to place the decimal point in each answer.

  a. 72.12 * 84.75 = 6 1 1 2 1 7

  b. 60.48 ÷ 3.2 = 1 8 9

  SRB
  128, 134,
  137

2 Johanna has $\frac{1}{4}$ gallon of water. She wants to divide it into equal amounts to drink after each mile of a 4-mile run. How much water should she drink after each mile?

  _____
  (number model)

  Answer: _____

  SRB
  207-208

3 What is the area of a sheet of legal-size paper that is $8\frac{1}{2}$ in. by 14 in.?

  _____
  (number model)

  Answer: _____

  SRB
  204-205
  225

4 Fill in the missing exponents to make true number sentences.

  a. $5.02 * 10^{\boxed{}} = 5,020,000$

  b. $934.27 ÷ 10^{\boxed{}} = 9.3427$

  c. $81.09 * 10^{\boxed{}} = 81,090$

  SRB
  133, 136

5 The deli workers at the grocery store are keeping track of how much turkey they sell. The line plot shows the amount of turkey in each order for one day. Use the line plot to answer the questions.

  a. What was the total amount of turkey sold in orders of 1 lb or more?

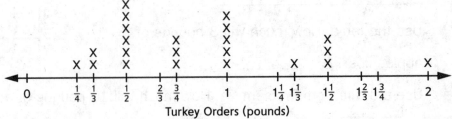

Turkey Orders (pounds)

  _____

  b. What was the total amount of turkey sold in orders of $\frac{1}{2}$ lb? _____

  SRB
  177, 191
  199, 247

Buying a Fish Tank

Math Message

SRB
225, 233

1. Glenn is buying a fish tank for his goldfish, Swimmy. Glenn learned that a 1-inch goldfish needs at least 230 cubic inches of water to be healthy. Swimmy is 1 inch long.

   Find the volume in cubic inches of each fish tank shown below.

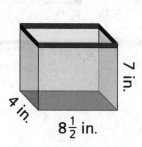

Fish Tank 1

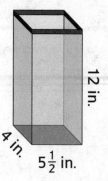

Fish Tank 2

Number model: $V =$ _____   Number model: $V =$ _____

Volume = _____   Volume = _____
            (unit)   (unit)

2. Which fish tank should Glenn buy for Swimmy? _____

   Explain your choice to a partner.

---

3. Glenn also learned that the area of the base of a fish tank should be at least 30 square inches for each inch of fish length.

   a. What is the area of the base of Fish Tank 1? _____
                                                         (unit)

   b. What is the area of the base of Fish Tank 2? _____
                                                         (unit)

   c. Which fish tank should Glenn buy for Swimmy? Explain.

   _____

   _____

Planning Your Aquarium

Imagine you are setting up an aquarium in your room. The pet store has four fish tanks available.

SRB
225, 227
233-234

Tank A: *Underwater Adventure*

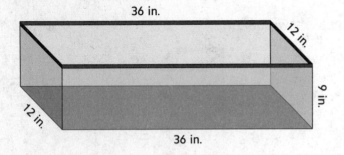

36 in.
12 in.
9 in.
12 in.
36 in.

Tank B: *Fish Hotel*

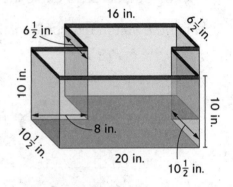

16 in.
$6\frac{1}{2}$ in.
$6\frac{1}{2}$ in.
10 in.
8 in.
$10\frac{1}{2}$ in.
10 in.
20 in.
$10\frac{1}{2}$ in.

Tank C: *Fish Palace*

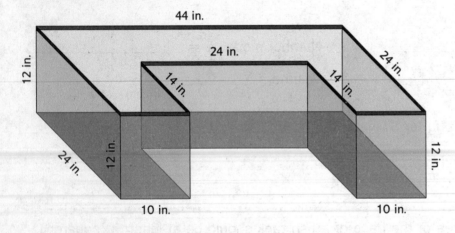

44 in.
24 in.
12 in.
14 in.
14 in.
24 in.
24 in.
12 in.
12 in.
10 in.
10 in.
12 in.

Tank D: *Water World*

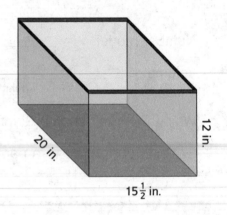

20 in.
12 in.
$15\frac{1}{2}$ in.

1. Choose the fish tank that you want for your aquarium. _____
   (name of fish tank)

2. What is the area of the base of your fish tank?

   _____
   (number model)

   A = _____
           (unit)

3. What is the volume of your fish tank?

   _____
   (number model)

   V = _____
           (unit)

Planning Your Aquarium (continued)

Keep in mind the following fish tank guidelines. For each inch of fish length:

- The tank must hold at least 230 cubic inches of water.

- The base of the tank should have an area of at least 30 square inches.

4   **a.** About _____ total inches of fish could live in my tank.

   b. Explain your thinking. _____

   _____

   _____

   _____

   _____

   _____

5   **a.** Complete the Goldfish Order Form on the next page.

   b. Explain how you know that the fish you chose will be healthy living in your fish tank.

   _____

   _____

   _____

Try This

6   Suppose you poured 10 gallons, or about 2,300 cubic inches, of water into your fish tank. About how high would the water level be in your tank? Use a calculator and round your answer to the nearest tenth of an inch. Show your work. *Hint:* Use the formula $V = B \times h$.

   The water level in my fish tank would be about _____ inches.

Goldfish Order Form

Goldfish Order Form			
Goldfish	Size	How Many?	Total Inches
Fantail goldfish	1-inch fantail		
	2-inch fantail		
	$2\frac{1}{2}$-inch fantail		
Lionhead goldfish	2-inch lionhead		
	3-inch lionhead		
	4-inch lionhead		
Black dragon eye goldfish	$1\frac{1}{2}$-inch black dragon eye		
	2-inch black dragon eye		
	$3\frac{1}{2}$-inch black dragon eye		

Combined length of all the fish: _____ in.

Reminder: Go back and complete the problems on the previous page.

Math Boxes

1 Choose the best answer.

When you multiply a number by a fraction greater than one, _____

○ the product is equal to the number.

○ the product is greater than the number.

○ the product is less than the number.

○ we don't have enough information to know about the size of the product.

SRB
197

2 Jacory made $223.20 for 12 hours of work. How much did he make in 1 hour?

_____
(number model)

Answer: _____

SRB
44, 137

3 Caetano is cutting out rectangles for an art project. Each rectangle has side lengths that are $\frac{7}{8}$ of the previous rectangle's side lengths. The last rectangle that Caetano cut out was $\frac{3}{4}$ in. by $\frac{2}{3}$ in. What will the side lengths of the next rectangle be?

_____ by _____

SRB
178,
201-203

4 A bakery sells pie slices that are $\frac{1}{6}$ of a pie. If they made 18 pies today, how many slices of pie will they have?

_____
(division number model)

Answer: _____

SRB
209-210

5 **Writing/Reasoning** Explain how you solved Problem 2.

_____

_____

_____

SRB
137

287

Juice Box

Clark has a juice box that is not completely filled with juice. The hole at the top of the box is open. Clark grabs the juice box and squeezes it, but the juice does not come out of the box.

1 What do you think happened to the shape of the juice box when Clark squeezed it?

2 What happened to the amount of juice in the box and the height of the juice when Clark squeezed it?

Math Boxes

1 Rosa solved the problem below.

$85.19 * 1.8 = 1,533.42$

Is Rosa's answer correct? _____

Write an estimate to show how you know:

_____

If the digits from Rosa's answer are correct, what is the product of $85.19 * 1.8$?

_____

SRB
128, 134

2 Mr. Havlis split $\frac{1}{3}$ gallon of paint equally among 10 students for a mural they were working on. How much paint did each student get?

_____
(number model)

Answer: _____ gallon

SRB
207-208

3 Brigitte is working on a quilt with squares of fabric measuring $\frac{7}{12}$ ft on each side. What is the area of each square?

_____
(number model)

Answer: _____

SRB
202-203
225

4 Fill in each blank with an exponent to make a true number sentence.

a. $4.3 * 10^{\boxed{}} = 43,000$

b. $8.7 \div 10^{\boxed{}} = 0.087$

SRB
133, 136

5 Phoebe has been tracking her height since she was 7 years old. Twice a year she records her growth from her previous measurement. Her measurements are given in inches below.

$$\frac{3}{4} \qquad 1\frac{1}{8} \qquad \frac{7}{8} \qquad \frac{1}{4} \qquad 1\frac{1}{2} \qquad \frac{3}{8} \qquad 1 \qquad 1\frac{1}{2} \qquad \frac{1}{8}$$

```
<---+-------+-------+-------+-------+-------+-------+-------+--->
    0      1/4     1/2     3/4      1     1 1/4   1 1/2   1 3/4    2
```

Phoebe's Growth (in.)

a. Plot the data on the line plot.

b. If Phoebe was $49\frac{5}{8}$ inches tall at age 7, how tall was she at her last measurement?

_____

SRB
177, 191
244, 247

Math Message

Imagine you are opening an animal shelter for stray dogs and cats. Talk with a partner about the supplies you think you would need. Make a list in the space provided.

How would you spend $1,000,000?

Your town had an essay contest called "How Would You Spend One Million Dollars?" You and your classmates submitted an essay about opening an animal shelter for stray cats and dogs, and you won! The town has given your class $1,000,000 to spend on the animal shelter.

As a class, you must develop a plan for spending the $1,000,000 to open and operate the shelter for one year. You have the following guidelines:

• The shelter needs to house up to 20 dogs and 120 cats.

• The cost of renting a building for the shelter is $8,000 per month.

• Water, heat, air conditioning, and electricity together cost approximately $450 per month.

• Phone and Internet service together cost about $200 per month.

• The town wants the shelter to create new full-time and part-time jobs. Veterinarians have to be paid at least $4,000 per month. Other employees have to be paid at least $1,800 per month.

Use the accounting sheets on the following pages to create your spending plan. Think about major categories to organize your spending. You will need to do some research to find the cost of individual items.

Major Categories Accounting Sheet

Major Category	Total Cost for Category

Total for all categories: _____

Itemized Accounting Sheet

Category: _____			
Item	Quantity	Unit Cost	Approximate Total Cost
Example: 6 lb bag of dog food	10	$19.99	10 * $20 = $200
Total for all categories: _____			

If the total is *under* budget, giving you more money to spend, how would you adjust your spending in this category? _____

If the total is *over* budget, requiring you to cut costs, how would you adjust your spending in this category? _____

Itemized Accounting
Sheet (continued)

Category: _____			
Item	Quantity	Unit Cost	Approximate Total Cost
Example: 6 lb bag of dog food	10	$19.99	10 * $20 = $200
Total for all categories: _____			

If the total is *under* budget, giving you more money to spend, how would you adjust your

spending in this category? _____

If the total is *over* budget, requiring you to cut costs, how would you adjust your spending in

this category? _____

293

Math Boxes

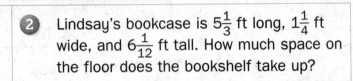

Math Boxes

① Divide.

 a. $\frac{1}{3} \div 6 =$ _____

 b. $6 \div \frac{1}{3} =$ _____

SRB 207-210

② Lindsay's bookcase is $5\frac{1}{3}$ ft long, $1\frac{1}{4}$ ft wide, and $6\frac{1}{12}$ ft tall. How much space on the floor does the bookshelf take up?

_____
    (number model)

Answer: _____

SRB 204-206 225

③ Ed used $1\frac{1}{4}$ dishes of paint on a project. Each dish holds $\frac{3}{5}$ fluid ounce of paint. How many fluid ounces did he use?

_____
    (number model)

Answer: _____

SRB 204-206

④ Multiply.

 a. $2.3 * 8.4 =$ _____

 b. $0.47 * 56.3 =$ _____

SRB 134-135

⑤ **Writing/Reasoning** How is multiplying decimals similar to multiplying whole numbers? How is it different?

_____

_____

_____

_____

_____

_____

SRB 134-135

294

Earning $1,000,000

Record the hourly wage assigned to your group: _____

How long would it take you to earn $1,000,000 at this hourly wage? Find the total amount of time you would need to work. Express your answer in two ways:

• As a total number of hours

• Using the largest possible units (You will need to use a combination of work years, workweeks, workdays, and hours.)

Show your work.

Total number of hours: _____

Answer in largest possible units: _____

_____

Multiplying Mixed Numbers

Solve Problems 1 and 2. Show your work.

1 $8\frac{1}{6} * 4\frac{2}{3} = ?$

2 $12\frac{3}{10} * 3\frac{5}{8} = ?$

SRB
204-206

$8\frac{1}{6} * 4\frac{2}{3} =$ _____

$12\frac{3}{10} * 3\frac{5}{8} =$ _____

For Problems 3 and 4, write a number model using a letter for the unknown. Then solve.
Show your work.

3 Leah is designing a poster for Math Night. She made a sketch of a Math Night logo that is $1\frac{3}{4}$ inches wide. She wants to make the logo $5\frac{1}{2}$ times as wide for the poster. How wide will the logo be on the poster?

4 Martin is helping his parents order new carpeting for their living room. The floor is a $14\frac{2}{5}$ ft by $18\frac{1}{3}$ ft rectangle. How many square feet of carpeting should Martin's family order?

_____
(number model)

_____
(number model)

Answer: _____

Answer: _____

5 What method did you use to solve Problem 1? Why did you choose that method?

_____

_____

Math Boxes

1 List three attributes of a square.

_____

_____

_____

How is a square different from a rectangle?

_____

_____

_____

SRB 269

2 Demetra is going to Greece and wants to exchange her money. The bank gives 0.73 euro for every U.S. dollar. How many euros will Demetra get if she exchanges 125 dollars?

_____
            (number model)

SRB 44, 134-135

Answer: _____

3 **a.** Use the rules to fill in the columns.

in (x) Rule: + 2	out (y) Rule: + 4
0	0

b. What rule relates each *in* number to its

corresponding *out* number? _____

c. Write the numbers from the table as ordered pairs. Then graph the ordered pairs and draw a line to connect the points.

Ordered pairs: _____ _____ _____

_____ _____

d. What is the *out* number when the *in* number is 7? _____

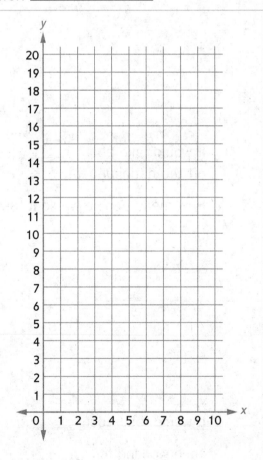

SRB 51-52 55, 275

297

Paying Off the National Debt

SRB
331-332
336

1. What is the current national debt? About _____

2. Record the hourly wage assigned to your group: _____

3. Record your estimate from the Math Message for the number of hours a typical person works in a lifetime: _____

Use the information above to solve Problem 4. You may use a calculator to help you, but be sure to record number models to keep track of your calculations. Check your work to make sure your answers are reasonable.

4. **a.** How many total work hours would it take to pay off the national debt at your assigned hourly wage?

   It would take about _____ hours to earn enough money to pay off the national debt.

   b. How many people would it take to work that many hours?

   It would take about _____ people earning _____ per hour and

   each working for _____ hours in their lifetime to earn enough money to pay off the national debt.

---

1 **a.** $7 \div \frac{1}{5} =$ _____

b. $\frac{1}{5} \div 7 =$ _____

SRB
207-210

---

2 Alma's bulletin board is $\frac{1}{3}$ as tall as it is wide. If the bulletin board is $2\frac{1}{4}$ m wide, how tall is it?

_____
(number model)

Answer: _____

SRB
204-206

---

3 Keith spent $2\frac{1}{2}$ hours doing chores. He spent $\frac{1}{4}$ of that time dusting. How much time did Keith spend dusting?

_____
(number model)

Answer: _____

SRB
204-206

---

4 Multiply.

a. $40.5 * 11.05 =$ _____

b. $0.95 * 20.2 =$ _____

SRB
134-135

---

5 **Writing/Reasoning** What is your favorite method for multiplying mixed numbers? Why do you like that method?

_____

_____

_____

_____

SRB
204-206

299

Footstep Problem

Imagine that you are living in a time when there are no cars, trains, or planes. You do not own a horse, a boat, or any other means of transportation.

SRB
215-217
328,332

You plan to travel to _____. You will have to walk there.

(location given by your teacher)

Information needed to solve the problem.

1. About how long is one footstep? About _____ feet

2. How many feet are in one mile? _____ feet

3. About how many miles is it from your school to your destination?

   About _____ miles

4. About how many steps does a fifth grader take in 1 minute?

   About _____ steps

5. a. About how many footsteps will you have to take to get from your school to your destination?

      About _____ footsteps

   b. Explain how you figured out about how many footsteps you would take.

      _____

      _____

      _____

      _____

      _____

SRB
215-217
328, 332

6 **a.** Suppose that you did not rest, eat, sleep, or stop for any reason. About how many hours would it take you to walk from school to your destination?

About _____ hours

b. Explain how you figured out about how many hours you would have to walk.

_____

_____

_____

_____

_____

Try This

7 Suppose you start from school at 7:00 A.M. on Monday. You do not rest, eat, sleep, or stop for any reason. On what day of the week, and at about what time, would you expect to reach your destination?

Day: _____

Time: About _____

Explain how you found your answer.

_____

_____

_____

_____

Solving Candle Problems

SRB
244, 248

Miranda had a set of ten identical candles. She put them around her house and burned them for different amounts of time. Now all of the candles are different heights.

She measured the height of each candle to the nearest $\frac{1}{8}$ inch. Here is her data:

Candle Heights (inches)				
$6\frac{1}{2}$	$6\frac{1}{4}$	$6\frac{1}{2}$	$6\frac{5}{8}$	$8\frac{1}{2}$
$7\frac{7}{8}$	$8\frac{1}{2}$	$7\frac{3}{8}$	$6\frac{1}{8}$	$6\frac{1}{4}$

1. Plot the candle height data on the line plot below.

Candle Heights

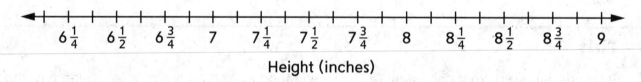

Height (inches)

2. How much taller is the tallest candle than the shortest candle?

Number model: _____

Answer: _____ inches taller

3. If the three shortest candles were stacked end to end, how high would the stack be?

Number model: _____

Answer: _____ inches high

4. Miranda is planning to melt all the remaining wax from the candles and then use it to make 10 candles that are the same width as the original candles and are all the same height. How tall will each of the 10 candles be?

Each candle will be _____ inches tall.

Math Boxes

Math Boxes

1 How many inches in a foot?

_____ inches

How many feet in a yard?

_____ feet

How many inches in a yard?

_____ inches

SRB
328

2 Justis is using $\frac{1}{3}$ cup of chopped walnuts for a muffin recipe. If the walnuts are evenly split among 8 muffins, how many cups of walnuts are in each muffin?

_____
(number model)

Answer: _____

SRB
207-208
210

3 Draw a parallelogram.

List two other names for your shape.

_____

_____

SRB
269

4 Giselle is making friendship necklaces and needs 0.25 m of string per necklace. How much string will she need to make necklaces for her 4 best friends?

_____
(number model)

Answer: _____

SRB
44,
134-135

5 **Writing/Reasoning** Think of a shape that would *not* be a correct answer for Problem 3. Draw it and explain why it is not a correct answer.

_____

_____

SRB
269

303

The Heart

Math Message

The heart is an organ in your body that pumps blood
through your blood vessels. Your **heart rate** is the number
of times your heart beats in a given amount of time. It
is usually expressed as heartbeats per minute. With
each heartbeat, your arteries stretch and then return
to their original size. This throbbing of your arteries is
called your **pulse.**

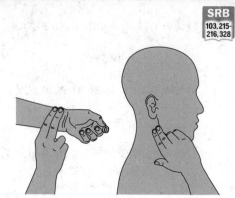

You can feel your pulse on the bottom side of your wrist,
below your thumb. You can also feel it in your neck.
Run your index and middle fingers from your ear past
the curve of your jaw, and press them gently into your neck just below your jaw.

1. Work with a partner. Find your pulse. Count the number of heartbeats in 15 seconds while
   your partner keeps track of the time. Do this several times until you are sure your count is
   accurate. Then switch roles. Record your results below.

   In 15 seconds my heart beats about _____ times.

2. How could you use your answer to Problem 1 to find out how many times your heart beats
   in 1 minute? Work with a partner. Show all of your work below.

   In 1 minute my heart beats about _____ times.

3. **a.** Fill in the blanks.

   1 hour = _____ minutes

   1 day = _____ hours

   1 year = _____ days

   b. Use multiplication and the unit
   conversions in Part a to help you
   complete the table below.

Time	Number of Heartbeats
1 minute	
1 hour	
1 day	
1 year	

Exercising Your Heart

Work with a partner to find out how exercise affects your heart rate.

1 Use your answers to Problems 1 and 2 on journal page 304 to fill in the first row of the table at the right. The rate at which your heart beats when you are sitting quietly is called your *resting heart rate*.

2 In an open area, do 10 jumping jacks without stopping. As soon as you finish, take your pulse for 15 seconds while your partner times you. Record the number of heartbeats in the second row of the table.

Number of Jumping Jacks	Heartbeats in 15 Seconds	Heartbeats in 1 Minute
0		
10		
20		
30		
40		
50		

3 Sit quietly. While you are resting, your partner can do 10 jumping jacks and you can time your partner.

4 When your pulse has slowed down to your resting heart rate, do 20 jumping jacks without stopping. As soon as you finish, take your pulse. Record the number of heartbeats in 15 seconds in the third row of the table. Then rest while your partner does 20 jumping jacks.

5 Repeat the procedure for 30, 40, and 50 jumping jacks.

6 Use the data you recorded to figure out how many times your heart beats in 1 minute after each number of jumping jacks. Record your answers in the third column of the table.

7 Why is it important for all students to do jumping jacks at about the same speed?

_____

_____

_____

8 Look at your table. Does exercise increase or decrease your heart rate? How do you know?

_____

_____

9 Do you think there is a rule you could use to predict the number of times your heart would beat in 15 seconds after you did 100 jumping jacks? Explain your answer.

_____

_____

_____

Finding the Area of Rectangles

Solve each problem. Pay close attention to the units of the side lengths and the area in each problem. You may need to convert one or more units. Write a number sentence to show how you solved.

SRB
202-203
225, 328

1

2 ft

$\frac{2}{3}$ yd

Area: _____ yd²

_____
(number sentence)

Area: _____ ft²

_____
(number sentence)

2

$\frac{7}{12}$ ft

6 in.

Area: _____ ft²

_____
(number sentence)

3 A park measures $\frac{1}{3}$ mile on one side and 1,320 feet in the other direction. What is the area of the park in square miles?

4 Bruce's stamp collection includes a stamp that is a $\frac{7}{8}$-inch square. What is the area of the stamp?

Area: _____ mi²

_____
(number sentence)

Area: _____ in.²

_____
(number sentence)

5 The bottom surface of a box measures $\frac{7}{10}$ m by $\frac{3}{10}$ m. What is the area in:

a. square meters? _____

b. square centimeters? _____

c. Explain how you found the area of the bottom of the box in square centimeters.

_____

_____

_____

306

Math Boxes

---

1 Solve. Use the common denominator method.

    a. $\frac{1}{5} \div 6 =$ _____

    b. $6 \div \frac{1}{5} =$ _____

SRB
210

---

2 Divide.

    $97.2 \div 8.1 = ?$

$97.2 \div 8.1 =$ _____

SRB
140-141

---

3   **a.** Multiply 80.73 by 10^2.

    _____

  b. Multiply your product from Part a

    by 10^3. _____

  c. How many total places did the decimal point move from 80.73 to your final

    product? _____

SRB
133

---

4 A rug has an area of $\frac{6}{15}$ m². Which of the following could be the dimensions of the rug? Circle ALL that apply.

  A. $\frac{3}{5}$ m by $\frac{2}{3}$ m

  B. $\frac{2}{15}$ m by $\frac{3}{15}$ m

  C. $\frac{5}{10}$ m by $\frac{1}{5}$ m

  D. $1\frac{1}{5}$ m by $\frac{1}{3}$ m

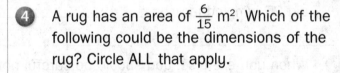
SRB
203-206

---

5 **Writing/Reasoning** Write a number story to match Problem 1a.

_____

_____

_____

SRB
207-208

My Heart-Rate Profile

Math Message

1 Write the data from the table on journal page 305 as ordered pairs. Use the numbers of jumping jacks as the *x*-coordinates and the numbers of heartbeats in 15 seconds as the *y*-coordinates. Then graph the points on the grid. Use a straightedge to connect the points in order.

Ordered pairs:

_____

_____

_____

_____

_____

_____

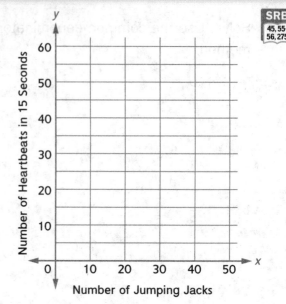

SRB
45, 55-
56, 275

2 Use the graph to predict how many times your heart would beat in 15 seconds after you

did 25 jumping jacks. _____

---

3 When you exercise, you should be careful not to put too much stress on your heart. Exercise experts often recommend a *target heart rate* for exercise. A person's target heart rate varies depending on the person's age and health, but the following rule is sometimes used to find a target heart rate in beats per minute:

Subtract your age from 220. Multiply the result by 2. Then divide by 3.

a. Write a number model with a letter to show how you could use this rule to find your target heart rate.

_____
(number model)

b. What is your target heart rate? _____ beats per minute

c. That is equal to how many heartbeats in 15 seconds? _____ beats in 15 seconds

d. Did you reach your target heart rate while you did jumping jacks? Explain how you know.

_____

_____

Cardiac Output

The amount of blood your heart pumps in 1 minute is called your **cardiac output.**
To find your cardiac output, you can multiply your heart rate by the amount of blood
your heart pumps with each heartbeat.

SRB
108-109
134, 217

Cardiac output = heart rate * amount of blood pumped with each heartbeat

Because cardiac output depends on a person's heart rate, it is not the same at all times.
The more the heart beats in 1 minute, the more blood is pumped throughout the body.

Follow these directions to find your cardiac output before and after exercise.

1. Look at the data on journal page 305. In the third column of the table below, write
   the number of times your heart beat in 15 seconds after 0 jumping jacks and after
   50 jumping jacks.

2. Use your answers to Problem 1 to figure out how many times your heart would beat in
   1 minute after 0 jumping jacks and after 50 jumping jacks. This is your heart rate before
   and after exercise. Write your answers in the fourth column.

3. The heart of a typical fifth grader pumps about 1.6 fluid ounces of blood with each
   heartbeat. Use this information to find your cardiac output before and after exercise.
   Record your cardiac output in the last column.

Before or After Exercise?	Number of Jumping Jacks	Heartbeats in 15 Seconds	Heartbeats in 1 Minute (Heart Rate)	Cardiac Output (fluid ounces per minute)
Before	0			
After	50			

4. a. Fill in the blank: 1 day = _____ minutes

   b. If you did not exercise, about how many fluid ounces of blood would your heart pump
      in 1 day?

      _____   About _____ fluid ounces
      (number model)

   c. Fill in the blank: 1 gallon = _____ fluid ounces

   d. If you did not exercise, about how many gallons of blood would your heart pump
      in 1 day?

      _____   About _____ gallons
      (number model)

309

Math Boxes

Math Boxes

1 Draw a quadrilateral with exactly one pair of parallel sides.

What is the most specific name for the

shape you drew? _____

SRB
269

2 Claire bought 1.35 lb of yellow onions for $0.80 per pound. How much did she spend on yellow onions?

_____
(number model)

Answer: _____

SRB
44,
134-135

3 **a.** Use the rules to complete the columns.

in (x)	out (y)
Rule: + 3	**Rule: + 3**
3	1

b. What rule relates each *in* number to its corresponding *out* number?

_____

c. Write the numbers from the table as ordered pairs. Then graph the ordered pairs and draw a line to connect the points.

Ordered pairs: _____ _____

_____ _____

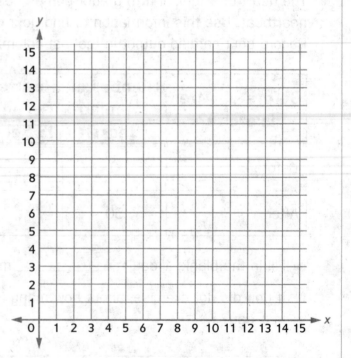

d. What is the *in* number when the *out* number is 16? _____

SRB
51-52
55, 275

Pendulums

A **pendulum** consists of an object called a **bob** suspended from a fixed support so that the bob can swing freely back and forth. You might have seen a pendulum on a clock.

According to legend, the Italian scientist Galileo began investigating pendulums in 1583 after he watched a hanging lamp swing back and forth in a cathedral in Pisa. Galileo discovered that each swing of a pendulum takes the same amount of time, so he began using pendulums as timing devices. Later his findings led Dutch mathematician Christiaan Huygens to invent the first pendulum clock.

Imagine that you have two pendulums, one longer than the other. Do you think the two pendulums would take the same amount of time to swing back and forth, or would one have a longer swing time than the other? Explain your answer.

pendulum

Clock with pendulum

_____

_____

_____

_____

_____

Investigating Pendulums

Your teacher will demonstrate an experiment with a pendulum that is 50 centimeters long.

SRB
126, 136

1 Record the time it took for the pendulum to complete 10 swings in the 50 cm row of the table at the bottom of the page.

2 Divide the time from Problem 1 by 10 to find about how long it took the pendulum to complete 1 swing. Round your answer to the nearest tenth of a second. Record the result in the table.

3 Describe a quick way that you can divide by 10 to help you complete the last column of the table. *Hint:* Think about place value.

_____

_____

_____

_____

4 Why might it be more accurate to time 10 swings and divide by 10 than to time just 1 swing?

_____

_____

Length of Pendulum	Time to Complete 10 Swings	Time to Complete 1 Swing
5 cm	_____ sec	_____ sec
10 cm	_____ sec	_____ sec
20 cm	_____ sec	_____ sec
30 cm	_____ sec	_____ sec
50 cm	_____ sec	_____ sec
75 cm	_____ sec	_____ sec
100 cm	_____ sec	_____ sec
200 cm	_____ sec	_____ sec

5 Write your data from the table on journal page 312 as ordered pairs. Use the pendulum length (in centimeters) as the *x*-coordinate. Use the time to complete *1 swing* (in seconds) as the *y*-coordinate.

For example, if your 5 cm pendulum took 0.5 second to complete 1 swing, your first ordered pair would be (5, 0.5).

Record your ordered pairs below.

_____   _____   _____   _____

_____   _____   _____   _____

6 Plot the points from Problem 5 on the grid below. Use line segments to connect the points in order.

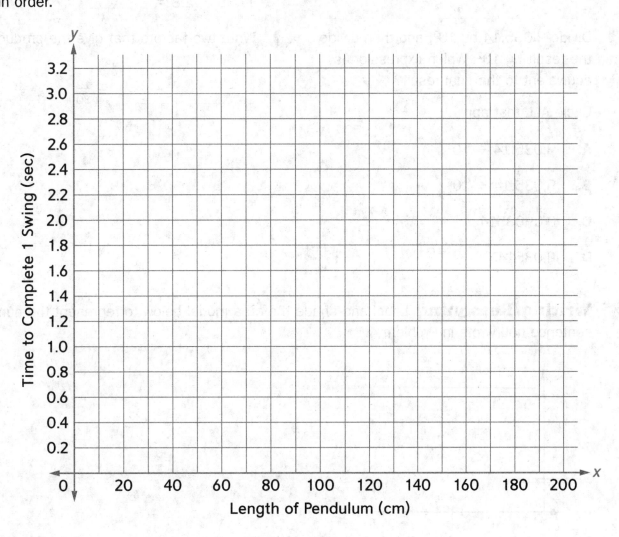

Math Boxes

(1) Solve. Use the common denominator method.

a. $\frac{1}{4} \div 12 = $ _____

b. $12 \div \frac{1}{4} = $ _____

SRB
210

(2) Divide.

$17.36 \div 2.8 = ?$

$17.36 \div 2.8 = $ _____

SRB
140-141

(3) Divide 9,035.14 by 10^3, and then divide the result by 10^2. Which expression is equivalent to the final result?

Circle ALL that apply.

A. $9,035.14 \div 10^5$

B. $9,035.14 \div 10^6$

C. 0.0903514

D. 9.03514

SRB
136

(4) Write two factors that give the product $\frac{12}{30}$.

_____ * _____ = $\frac{12}{30}$

SRB
203

(5) **Writing/Reasoning** Label and shade the area model below to represent the number sentence you wrote in Problem 4.

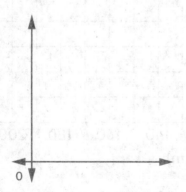

SRB
202

314

Using a Hierarchy

Use the Quadrilateral Hierarchy Poster or journal page 251 to help you answer the questions.

① Fill in the blanks with a subcategory that makes the statement true. Use the same subcategory in both blanks.

All _____ are kites, but not all kites are _____.

② **a.** Draw a trapezoid that is also a parallelogram.

b. Draw a trapezoid that is *not* a parallelogram.

c. Explain how you decided what to draw for Part b.

_____

_____

_____

③ **a.** Classify the shape at the right on the hierarchy.
List all of the categories you passed through while you were classifying.

_____

_____

b. Explain how you knew when to stop classifying the quadrilateral. How did you know that the shape couldn't move down into more subcategories?

_____

_____

_____

④ **a.** Draw a quadrilateral with fewer than 3 names.

b. Draw a quadrilateral with exactly 3 names.

c. Draw a quadrilateral with more than 3 names.

Investigating Arc Size

Math Message

SRB
55-56
136, 275

In Lesson 8-11 you discovered that longer pendulums have longer swing times. For Problems 1–3, use the graph on journal page 313 to make predictions about swing times of pendulums of different lengths.

1 To the nearest tenth of a second, how long do you think it would take a 150 cm pendulum to complete 1 swing? About _____ seconds

2 To the nearest 5 centimeters, what length do you think a pendulum that takes 2 seconds to complete 1 swing would be? About _____ centimeters in length

3 Can you use this graph to predict how long it would take a 300 cm pendulum to complete 1 swing? Explain why or why not.

_____

_____

_____

4 Do you think that arc size will affect the time it takes a pendulum to complete 1 swing? Explain why or why not.

_____

_____

5 With your small group, make a 50 cm pendulum. Time how long it takes for the pendulum to complete 10 swings for each of the arc sizes in the table. Record your results in the table on journal page 317.

Remember: 30° is about one-third of the way to horizontal, 45° is about halfway, 60° is about two-thirds of the way, and 90° means that the string is horizontal (parallel to the floor).

6 Divide each 10-swing time by 10 to find a 1-swing time for each arc size. Round the times to the nearest tenth of a second. Record your results in the table on journal page 317.

7 Examine your completed table. Does arc size affect swing time? Explain how you know.

_____

_____

8 Write your data from the table as ordered pairs. Use the arc size (in degrees) as the x-coordinate. Use the time to complete 1 swing (in seconds) as the y-coordinate.

Arc Size	Time to Complete 10 Swings	Time to Complete 1 Swing
about 30°	_____ sec	_____ sec
about 45°	_____ sec	_____ sec
about 60°	_____ sec	_____ sec
about 90°	_____ sec	_____ sec

Record your ordered pairs below.

_____

_____

_____

_____

9 Plot the points from Problem 8 on the grid below. Use line segments to connect the points in order.

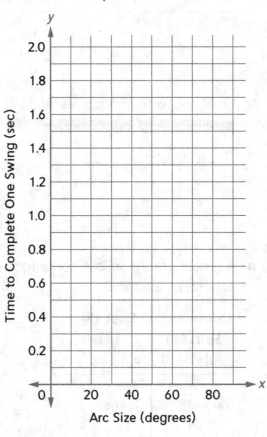

10 How long do you think it would take for a 50 cm pendulum to complete 1 swing if the swing had a 50-degree arc size? About _____ seconds

11 What does your completed graph show about the effect of arc size on swing time?

_____

_____

_____

12 Compare the graph above with the graph on journal page 313. How are the graphs different?

_____

_____

_____

Math Boxes

1 **a.** What property do squares and rhombuses both have?

_____

b. What property do squares and parallelograms both have?

_____

SRB
269

2 A kilogram is about 2.2 pounds. What is the weight in pounds of someone who weighs 49.6 kilograms?

_____
(number model)

Answer: About _____

SRB
44, 134-
135, 216

3 **a.** A grocery store sells 4 containers of yogurt for $2. Use the rule to complete each column of the table.

Containers Sold (x) Rule: + 4	Cost ($) (y) Rule: + 2
0	0

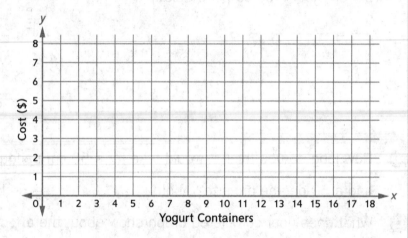

Cost ($)

Yogurt Containers

b. What rule relates each x value to its corresponding y value?

_____

c. Write the numbers from the table as ordered pairs. Then graph the ordered pairs and draw a line to connect the points.

Ordered pairs: _____ _____ _____ _____ _____

d. What is the cost of 1 yogurt container? $_____

SRB
51-52
55, 275

Math Boxes

1 1 foot = _____ inches

1 mile = 5,280 feet

How many inches are in 1 mile?

Answer: _____ inches   [SRB 328]

2 Rolando has collected $\frac{1}{2}$ truckload of books to donate to the library. If 6 people equally share the work of unloading the truck, what part of the truckload is each person responsible for?

_____
(number model)

Answer: _____   [SRB 207-208 210]

3 Draw a 4-sided figure with no pairs of parallel sides.

Give one name for the shape you drew.

_____   [SRB 269]

4 One kilometer is about 0.62 mile. A marathon is 26.2 miles. About how many kilometers is a marathon?

_____
(number model)

Answer: About _____ kilometers   [SRB 44 140-141]

5 **Writing/Reasoning** Explain how you solved Problem 2.

_____

_____

_____

[SRB 207-208 210]

Fraction Of Fraction Cards (Set 2)

$\frac{2}{3}$ | $\frac{2}{4}$ | $\frac{3}{4}$ | $\frac{2}{5}$

$\frac{3}{5}$ | $\frac{4}{5}$ | $\frac{2}{10}$ | $\frac{3}{10}$

$\frac{4}{10}$ | $\frac{5}{10}$ | $\frac{6}{10}$ | $\frac{7}{10}$

$\frac{8}{10}$ | $\frac{9}{10}$ | $\frac{0}{10}$ | $\frac{4}{4}$

0.22 _____
second

0.21 _____

0.20 _____

0.19 _____

0.18 _____

0.17 _____

0.16 _____

0.15 _____

0.14 _____

0.13 _____

0.12 _____

0.11 _____

0.10 _____

0.09 _____

0.08 _____

0.07 _____

0.00 **starting position** _____
     for contestant

Spoon Scramble Cards

✂️

$\frac{1}{4}$ of 24	$\frac{3}{4} * 8$	$6{,}000 \div 10^3$	$0.06 * 10^2$
$\frac{1}{3}$ of 21	$3\frac{1}{2} * 2$	$0.01 * 700$	$0.007 * 10^3$
$\frac{1}{5}$ of 40	$2 * \frac{16}{4}$	$80{,}000 \div 10^4$	$0.8 * 10^1$
$\frac{3}{4}$ of 12	$4\frac{1}{2} * 2$	$9 \div 10^0$	$0.0009 * 10^4$

Property Pandemonium Cards

Property Card at least 1 pair of parallel sides	Property Card 2 pairs of parallel sides	Property Card 2 pairs of adjacent sides equal in length	Property Card 4 sides equal in length
Property Card 4 right angles	Property Card 2 pairs of parallel sides and 4 right angles	Property Card at least 1 pair of parallel sides and 4 sides equal in length	Property Card WILD
Quadrilateral Card trapezoid	Quadrilateral Card parallelogram	Quadrilateral Card rhombus	Quadrilateral Card rectangle
Quadrilateral Card kite	Quadrilateral Card quadrilateral	Quadrilateral Card square	Quadrilateral Card WILD

Blank Fraction Circles

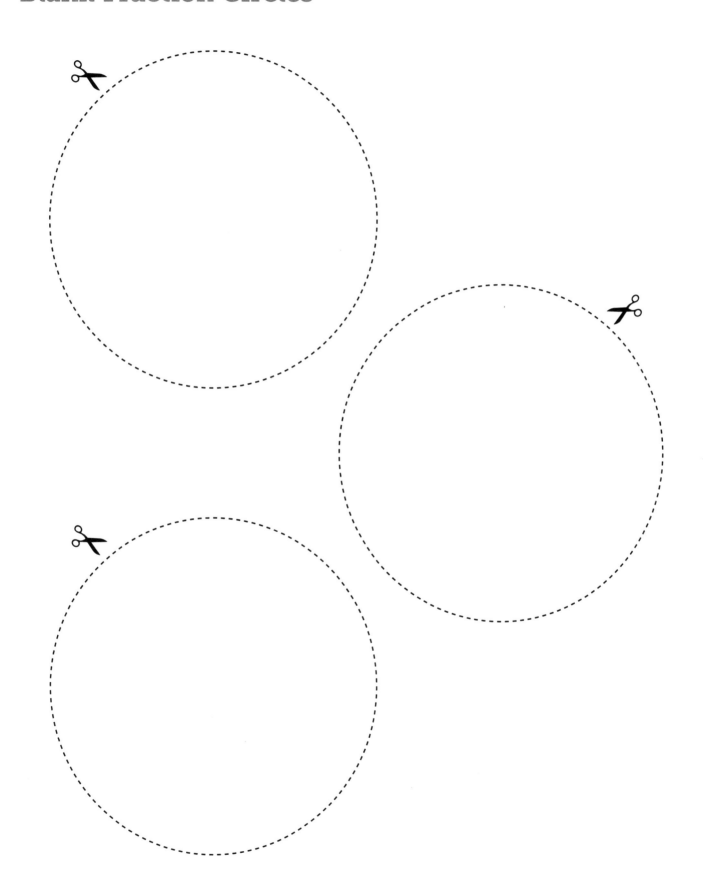

Blank Cards

Blank Cards

Blank Cards